The Kuhnian Image of Science

Collective Studies in Knowledge and Society

Series Editor: James H. Collier is Associate Professor of Science and Technology in Society at Virginia Tech.

This is an interdisciplinary series published in collaboration with the Social Epistemology Review and Reply Collective. It addresses questions arising from understanding knowledge as constituted by, and constitutive of, existing, dynamic and governable social relations.

The Kuhnian Image of Science

Time for a Decisive Transformation?

Edited by Moti Mizrahi

ROWMAN &
LITTLEFIELD
INTERNATIONAL

London • New York

Published by Rowman & Littlefield International, Ltd.
6 Tinworth Street, London SE11 5AL, United Kingdom
www.rowmaninternational.com

Rowman & Littlefield International Ltd. is an affiliate of Rowman & Littlefield
4501 Forbes Boulevard, Suite 200, Lanham, Maryland 20706, USA
With additional offices in Boulder, New York, Toronto (Canada), and Plymouth (UK)
www.rowman.com

British Library Cataloguing in Publication Data
A catalogue record for this book is available from the British Library

ISBN: HB 978-1-78660-340-1
ISBN: PB 978-1-78660-341-8

Library of Congress Cataloging-in-Publication Data Is Available
ISBN 978-1-78660-340-1 (cloth: alk. paper)
ISBN 978-1-78660-341-8 (paperback)
ISBN 978-1-78660-342-5 (electronic)

∞™ The paper used in this publication meets the minimum requirements of American
National Standard for Information Sciences—Permanence of Paper for Printed Library
Materials, ANSI/NISO Z39.48–1992.

Contents

Introduction

Moti Mizrahi

1. QUESTIONING THE KUHNIAN IMAGE OF SCIENCE

On the occasion of the fiftieth anniversary of the publication of Thomas Kuhn's *The Structure of Scientific Revolutions* (1962), *The Guardian* published an article entitled "Thomas Kuhn: The Man Who Changed the Way the World Looks at Science" (Naughton 2012). Kuhn's influence is undeniable. His way of looking at science remains enormously influential to this day. This is evidenced by the following:

- Kuhn's *The Structure of Scientific Revolutions* (1962) is the most cited book in the social sciences with 81,311 citations and counting (Green 2016).
- Many books (see, e.g., Wray 2011), edited volumes (see, e.g., Kindi and Arabatzis 2012), and special journal issues (see, e.g., Levine and Feingold 2010) are devoted to Kuhn's image of science.
- Historians, philosophers, and sociologists of science routinely use Kuhnian terminology to talk about science. Terms like "scientific revolution," "paradigm shift," and "incommensurability" are now common currency among scholars, scientists, and even laypeople.

In addition to the inclusion of "paradigm shift" in *The Oxford English Dictionary*, according to which the origin of the term is "the writings of Thomas S. Kuhn (1922–96)," the fact that Kuhnian terminology has spilled from academic discourse over to public discourse can be seen by looking at the topic "paradigm shift" in Google Trends. Figure 1.1 shows search interest in the topic relative to the highest point on the chart worldwide. A value of 100 is the peak popularity for the topic, as in March 2010. In 2010, at least three books on Kuhn's image of science were published (see D'Agostino 2010;

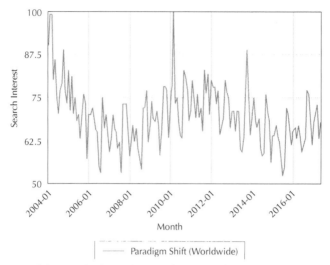

Figure 1.1. Search interest in the topic "paradigm shift" by month from January 2004 to January 2016 (Source: Google Trends).

Onkware 2010; and Torres 2010). Likewise, a value of 50 means that the topic is half as popular, as in July 2015.

All of this is not to say that Kuhn's image of science did not have its fair share of critics (see, e.g., Scheffler 1967). For some reason, however, more attention was paid to the details of Kuhn's image of science than to the *arguments for* it. Some even talk about Kuhn's image of science as a "discovery," thereby suggesting that science is, as a matter of fact, just the way Kuhn says it is (Hoyningen-Huene 1993, 41; Oberheim 2005; Oberheim and Hoyningen-Huene 2016). In Mizrahi (2015a), then, I set out to examine more closely the arguments for the Kuhnian image of science. After all, there must be pretty good arguments for such an influential philosophy of science.

What I have found, as I argue in Mizrahi (2015a), which is reprinted with permission in this volume (chapter 1 of part I), is that Kuhn does not offer much by way of epistemic support for his image of science in general and his incommensurability thesis in particular. That is to say, I could not find a valid deductive argument for the following Kuhnian thesis about theory change in science (Mizrahi 2015a, 362):

Taxonomic Incommensurability (TI): Periods of scientific change (in particular, revolutionary change) that exhibit TI are scientific developments in which existing concepts are replaced with new concepts that are *incompatible* with the older concepts. The new concepts are incompatible with the old concepts in the following sense: two competing scientific theories are conceptually incompatible (or incommensurable) just in case they do not share the same "lexical taxonomy." A lexical taxonomy contains the structures and vocabulary that are used to state a theory (See Kuhn 2000, pp. 14–15, 63, 92–97, 229–33, 238–39, and 242–44; cf. Sankey 1997).

There is no deductive support for (TI) because conceptual incompatibility does not necessarily follow from the fact that kind terms have changed or have been abandoned in a transition from one theoretical framework to another (Mizrahi 2015a, 363–68). In addition, there is no inductive support for (TI) because, even if there were a few cases in the history of science in which an old theory was replaced by a conceptually incompatible new theory, such isolated cases would not provide a strong inductive basis for drawing general conclusions about the nature of scientific change as a whole (Mizrahi 2015a, 368–73). Despite the fact that, in the introduction to *The Structure of Scientific Revolutions* (1962), Kuhn famously writes:

> History, if viewed as a repository for more than anecdote or chronology, could produce a decisive transformation in the image of science by which we are now possessed (Kuhn 1962/1996, 1),

Kuhn is guilty of doing precisely what he says philosophers of science should not do, namely, use the history of science as a repository for anecdotes.

If this is correct, then Kuhn's *The Structure of Scientific Revolutions* (1962) may have been patient zero for an infectious disease I call *anecdotiasis*. Patients who suffer from this infection exhibit the following symptom: they use cherry-picked anecdotes or case studies (from the history of science) abnormally to support general claims (about the nature of science as a whole). Since Kuhn was "one of the most influential philosophers of science of the twentieth century, perhaps the most influential" (Bird 2013), he was the ideal carrier and the disease has spread to almost all quarters of the field of philosophy of science. Soler et al. (2014, 42) point out that "Kuhn is often credited with having initiated an 'historical turn' in the Anglo-American philosophy of science" (see also Wray 2011, 87). Theorizing about science in a way that is informed by a careful and meticulous investigation of the history of science would have been a great idea. Unfortunately, the execution of this idea was less than great. If I am right, the so-called historical turn that Kuhn has initiated is nothing more than an "anecdotal turn." Rather than carefully study the history of science, Kuhn cherry-picked isolated case studies. Without a reason to think that these cherry-picked case studies are representative of science as a whole, the ground on which Kuhn built his image of science is unstable.

This general claim about the pathological state of the field of philosophy of science in general, and general philosophy of science in particular, is supported, not by anecdotes, but by the following statistical data. Of all the articles currently (May 23, 2017) available in the PhilSci-Archive, which is "an electronic archive specifically tailored to and run by philosophers of science" (philsci-archive.pitt.edu), 3,415 of the 5,518 (62%) contain the words "case study." There is even a "History of Science Case Studies" category on PhilSci-Archive, which contains more papers (263) than the other 33 categories in the "General Issues" subject, with the exception of the following five categories: "Causation" (444), "Confirmation/Induction" (376), "Explanation" (363), "Realism/Anti-realism" (392), and "Structure of Theories" (284).

Regarding the PhilSci-Archive's "Realism/Anti-realism" category in particular, the data suggest that philosophers of science frequently use cherry-picked anecdotes or case studies from the history of science as evidence in the realism/anti-realism debate in general philosophy of science. Almost half (49%) of the papers in this category contain the words "case study" (191 out of 392). Of these papers, 29% discuss the same case study, namely, the case of phlogiston (55 out of 191). To cite just one of those papers as an example, Ladyman (2011) uses the phlogiston case as support for ontic structural realism. It is important to note, however, that Ladyman's (2011) use of the phlogiston case as evidence for ontic structural realism is by no means unique in general philosophy of science and that other repositories of philosophy of science papers bear this out. A quick search through the "Scientific Realism" category in PhilPapers (philpapers.org) turns up quite a few papers with the words "case study" in the title. Here are just a few examples (in the order in which they appear in the search results):

- "Dirac's Prediction of the Positron: A Case Study for the Current Realism Debate" (Pashby 2012).
- "Chemical Atomism: A Case Study in Confirmation and Ontology" (Brown 2015).
- "Constructive Empiricism and Scientific Practice. A Case Study" (Iranzo 2002).
- "Sommerfeld's Atombau: A Case Study in Potential Truth Approximation" (Hettema and Kuipers 1995).
- "A Case Study in Realism: Why Econometrics Is Committed to Capacities" (Cartwright 1988).

To be clear, these examples are provided simply for context. The evidence for the claim about the spread of *anecdotiasis* within the field of philosophy of science in general, and general philosophy of science in particular, is the aforementioned statistical data from the PhilSci-Archive and PhilPapers.

For additional statistical evidence for my diagnosis of the pathological state of philosophy of science, I have looked at research articles published in the leading journal in the field, namely, *Philosophy of Science*, which contain the words "case study," relative to the total number of research articles published in the journal between 1934 and 2010. I have used JSTOR Data for Research (dfr.jstor.org) to access these data, which are summarized in figure 1.2.

As we can see from figure 1.2, after 1962, which is the year in which *The Structure of Scientific Revolutions* (1962) was originally published, there is actually a decline in the number of research articles published in the journal *Philosophy of Science* that contain the words "case study." So it looks like these data *do not* support my hypothesis that Kuhn was patient zero as far as the spread of *anecdotiasis* in philosophy of science is concerned. For if he were, we would expect to see an increase in the proportion of research articles that contain the words "case study" published in *Philosophy of Science* shortly after 1962. Nevertheless, we can see that research articles that contain the words "case study" do make up a significant proportion of research articles published in *Philosophy of Science*, with peak years (i.e., years in which research articles that contain the words "case study" count for more than 50% of total research articles published in *Philosophy of Science*) being 1944, 1945, 1946, 1951, 1952, 1958, 1960, 1984, 1989, 1991, 1994, 1998, 2004, 2006, 2007, and 2008.

Similar results can be obtained by running the same search for the words "case study" on data from another leading journal in the field, namely, the *British Journal for the Philosophy of Science (BJPS)* in JSTOR. That is, consistent with the aforementioned data from the journal *Philosophy of Science*, there is no evidence for the hypothesis that Kuhn was patient zero, but the symptoms of *anecdotiasis* are quite clear, since research articles that contain the words "case study" do make up a significant proportion of research articles published in the *BJPS*, with peak years (i.e., years in which research articles that contain the words "case study" count for more than 50% of total

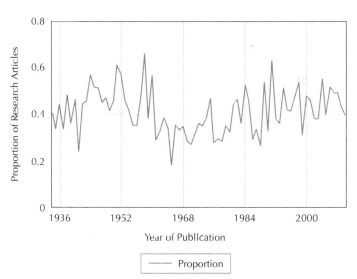

Figure 1.2. Proportions of research articles that contain the words "case study" published in the journal *Philosophy of Science* from 1936 until 2000 (Source: JSTOR Data for Research).

research articles published in the *BJPS*) being 1952, 1980, 1987, 1992, 1993, 1998, 2002, 2007, and 2010, as we can see from figure 1.3.

Since cherry-picking evidence, hasty generalization, and arguing from anecdotal evidence are all mistakes in inductive reasoning, it would be bad enough to find that *some* research articles in philosophy of science engage in such fallacious inductive reasoning. To find out, based on statistical data collected from PhilSci-Archive, PhilPapers, and JSTOR that a significant proportion of research articles published in the leading journals in the field, namely, *Philosophy of Science* and the *BJPS*, engage in such fallacious inductive reasoning should be concerning (see table 1.1).

My concern about the spread of *anecdotiasis* in the field of philosophy of science has prompted me to examine the arguments for Kuhn's image of science in great detail. Unfortunately, I have found that, despite proclaiming that history should not be used as a repository for anecdotes, Kuhn exhibits symptoms of *anecdotiasis* in *The Structure of Scientific Revolutions* (1962). In other words, I have found that Kuhn does precisely what he says philosophers

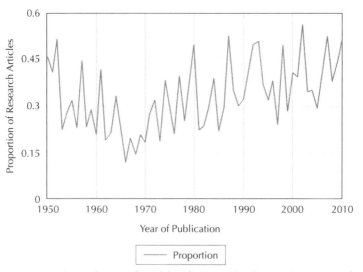

Figure 1.3. Proportions of research articles that contain the words "case study" published in the *BJPS* from 1950 until 2010 (Source: JSTOR Data for Research).

Table 1.1. Mean ratios of research articles that contain the words "case study" in the journals *Philosophy of Science* and the *BJPS*

	Mean	*SD*	*N*
Philosophy of Science	0.43	0.09	77
BJPS	0.34	0.11	61

of science should not do, namely, use the history of science as a repository for anecdotes. I argue that Kuhn uses—whether intentionally or not does not matter (Mizrahi 2015d, 133–34)—the history of science to cherry-pick anecdotes or case histories that seem to support his image of science, according to which, in periods of revolutionary change, theories are replaced with conceptually incompatible ones (Mizrahi 2015a). If I am right about this, then perhaps the most compelling argument Kuhn ever gave in support of his theory of scientific change is the "ashtray argument" (Morris 2011). And perhaps it is time for another "decisive transformation in the image of science by which we are now possessed." Only this time, the image of science that needs to be decisively transformed is the Kuhnian image of science.

Thanks to the executive editor of the journal *Social Epistemology*, James H. Collier, my close examination of the arguments (or lack thereof) for the Kuhnian image of science has led to a fruitful exchange on the *Social Epistemology Review and Reply Collective* (social-epistemology.com) with the philosophers of science Lydia Patton (2015), James Marcum (2015), and Vasso Kindi (2015), and my replies (Mizrahi 2015b; 2015c), and for which I am grateful. This volume was born out of this fruitful exchange, and it is designed to expand on it. The chapters in this volume critically examine the arguments for and against the Kuhnian image of science, as well as their implications for our understanding of science as a social and epistemic enterprise. Does the Kuhnian image of science provide an adequate model of scientific change? If we abandon the Kuhnian picture of revolutionary change and incommensurability, since there are neither epistemic nor pragmatic reasons to accept it (Mizrahi 2015a), what consequences would follow from that vis-à-vis our understanding of science as a social, epistemic endeavor?

By saying that science is a social and epistemic enterprise I mean that the social practices of scientists have epistemic effects. In other words, these social practices partly determine whether or not science can attain its epistemic goals. As Steve Fuller (1988/2002, 3) puts it:

The fundamental question of [. . .] social epistemology is: *How should the pursuit of knowledge be organized, given that under normal circumstances knowledge is pursued by many human beings, each working on a more or less well-defined body of knowledge and each equipped with roughly the same imperfect cognitive capacities, albeit with varying degrees of access to one another's activities?* (original emphasis)

On the Kuhnian image of science, scientists' "access to one another's activities" is further complicated (perhaps even rendered impossible) by translation failure and communication breakdown across "paradigms." This raises questions as to whether the Kuhnian image of science is the right model for understanding the social dimensions of science. For instance, as Helen Longino (2016) points out, "The second half of the twentieth century saw the emergence of what has come to be known as Big Science: the organization of large numbers of scientists bringing different bodies of expertise to a common research project." Can the Kuhnian image of science accommodate Big Science (e.g., large-scale scientific projects like the Human Genome Project and the Large Hadron Collider)? Can the Kuhnian image of science explain how a string theorist and an M-theorist can collaborate on the Compact Muon Solenoid experiment at the Large Hadron Collider in Geneva? After all, on the Kuhnian image of science, the string theorist and the M-theorist are supposed to be working in "different worlds" (Kuhn 1962/1996, 118, 150). The former works in a world of strings, whereas the latter works in a world of membranes.

In the second chapter of Part I of this volume, "Modeling Scientific Development: Lessons from Thomas Kuhn," Alexandra Argamakova also expresses concerns about "the validity of meta-theorizing and universal generalizations about scientific practice and history" (p. 46) and the ability of the Kuhnian image of science to accommodate interdisciplinary work in science. Argamakova argues that, if "paradigm" means a "disciplinary matrix," as Kuhn states in the Postscript to the second edition of *The Structure of Scientific Revolutions* (1962/1996, 182), then it is difficult to see how his model can explain the fact that science is becoming more interdisciplinary (Porter and Rafols 2009). Presumably, if disciplinary matrices were incommensurable, then practitioners from different disciplines would not be able to understand each other, let alone work together on shared projects, as in the case of the Human Genome Project and the Large Hadron Collider.

In addition to the worry about how the Kuhnian image of science can account for interdisciplinarity in science, Argamakova also points out that Kuhn's image of science is informed solely by the physical sciences, particularly physics, to the exclusion of the social and human sciences. As such, Argamakova argues, Kuhn's image of science cannot be a representative image of science as a whole. She finds the quest for a general theory of science to be "a peculiar ambition" (p. 57), given the diversity of scientific practices. If Argamakova is right, then there can be no general theory of scientific change along the lines of the Kuhnian image of science.

In chapter 3, "Can Kuhn's Taxonomic Incommensurability Be an Image of Science?" Seungbae Park finds more problems with the Kuhnian image of science. As Park (2011) himself, Ludwig Fahrbach (2011), and myself (Mizrahi 2013a, 2015d, and 2016) have pointed out, the selected case histories that are typically offered as support for "revolutions" in science are biased toward theories from before the early twentieth century. This is problematic because, as Fahrbach (2011, 148) observes, "half of all scientific work ever done was done in the last 15–20 years, while the other half was done in all the time before; and three quarters of all scientific work ever done was done in the last 30–40 years, while in all the time before that, only one quarter was done."

In addition to the worry that the Kuhnian image of science is based on fallacious inductive reasoning, Park finds Kuhn's appeal to evolutionary theory to be problematic as well. In particular, Park argues that to explain scientific change by analogy with evolutionary change is to presuppose that evolutionary theory is (approximately) true. But Kuhn can presuppose no such thing, for according to Kuhn's image of science, the paradigm that has dominated the life sciences since the publication of Charles Darwin's *On the Origin of Species* in 1859 will likely be superseded by a new paradigm that will be taxonomically incommensurable with it.

Worse still, according to the Kuhnian image of science, "scientific development isn't cumulative on the theoretical or on the factual level" (Preston 2008, 111). But evolution by natural selection is supposed to be cumulative (Sterelny and Griffiths 1999, 36). Accordingly, if evolution by natural selection is cumulative, but scientific evolution is not supposed to be cumulative, then it is difficult to see how the former can serve as a model for the latter.

In chapter 4, Howard Sankey proclaims "The Demise of the Incommensurability Thesis." Sankey argues against the claim that taxonomic incommensurability is a phenomenon that Kuhn "discovered" (see, e.g., Hoyningen-Huene and Oberheim 2009, 208). As the story goes, Kuhn "discovered" the taxonomic incommensurability of scientific theories

> as a graduate student in the mid to late 1940s while struggling with what appeared to be nonsensical passages in Aristotelian physics (Kuhn 2000 [1989], 59–60). He could not believe that someone as extraordinary as Aristotle could have written them. Eventually patterns in the disconcerting passages began to emerge, and then all at once, the text made sense to him: a Gestalt switch that resulted when he changed the meanings of some of the central terms (Oberheim and Hoyningen-Huene 2016).

As Sankey argues, however, the phenomenon described here is "the act of comprehension following the initial failure to understand the text of Aristotle" (p. 83). Kuhn proposes to explain this phenomenon by positing a relation of taxonomic incommensurability between scientific theories, in this case, Aristotelian physics and Newtonian physics. But there is no need to posit such a relation in order to explain the fact that Kuhn was able to understand a text that was initially incomprehensible to him. The only relation we need to posit here is that of understanding between a reader (namely, Kuhn) and a text. In general, comprehensibility and understanding are relations between a subject and a subject matter or a text, whereas incommensurability is supposed to be a relation between scientific theories or "paradigms." There is no straightforward entailment from the former to the latter. In Kuhn's case, the simplest explanation seems to be that he was eventually able to understand a text that he could not comprehend initially. There is no need to posit extra relations in order to explain that.

Sankey's remarks point to another problem with the Kuhnian image of science. Early critics of Kuhn accused him of confusing philosophy of science with psychology (Kuhn 1970), or the context of justification with the context of discovery (Hoyningen-Huene 2006). Whether or not the context of discovery/justification distinction is a real distinction, Sankey's remarks suggest that there is another sense in which Kuhn has confused philosophy of science with psychology. For, on the one hand, the question of theoretical change in science concerns scientific theories, and so taxonomic incommensurability is supposed to be a relation between theoretical frameworks. On the other hand, understanding is not a relation between theories; rather, it is a relation between a person and a theory. Two scientific theories, T_1 and T_2, can be taxonomically incommensurable, but T_1 cannot understand or fail to understand T_2. Likewise, a subject, S, can understand or fail to understand a scientific theory, T, but S cannot be taxonomically commensurable or incommensurable with T. If this is correct, then Kuhn is guilty of confusing semantic relations, which hold between theories, with psychological relations, which hold between subjects and theories. In Kuhn's case, his initial failure to understand Aristotelian physics, followed by his act of comprehension, is supposed to be evidence that Aristotelian physics and Newtonian physics are taxonomically incommensurable. But again, the former is a psychological relation between Kuhn and Aristotelian physics, whereas the latter is a semantic relation between Aristotelian physics and Newtonian physics. If Kuhn's "act of comprehension following the initial failure to understand the text of Aristotle" (p. 83) is supposed to be evidence that Aristotelian physics and Newtonian physics are taxonomically incommensurable, we need a plausible inferential bridge from psychological relations between subjects and theories to semantic relations between theories.

One of the reasons for this confusion between semantic relations and psychological relations may have something to do with the fact that Kuhn's forays into the history of science in *The Structure of Scientific Revolutions* (1962) are by and large instances of "Great Men" historiography. Evidence for this can be gleaned from the index of *The Structure of Scientific Revolutions* (1962/1996) (see table 1.2).

As we can see from table 1.2, the "Great Men" historiography takes up a significant proportion of the discussion in *The Structure of Scientific*

Table 1.2. The entries in the index for Kuhn's *The Structure of Scientific Revolutions* (1962/1996) divided into names of "Great Men" of science and the theoretical terms of Kuhn's image of science

	Names of "Great Men" of Science	Kuhnian Terminology
	Archimedes; Aristarchus; Aristotle; Sir Francis Bacon; J. Black; R. Boyle; Tycho Brahe; Copernicus; C. Coulomb; J. Dalton; C. Darwin; L. de Broglie; R. Descartes; A. Einstein; B. Franklin; Galileo Galilei; J. Hutton; Lord Kelvin; J. Kepler; A. Lavoisier; G. W. Leibniz; Sir Charles Lyell; J. C. Maxwell; Sir Isaac Newton; W. Pauli; M. Planck; J. Priestley; Ptolemy; W. Roentgen; C. Scheele	Anomalies; Crisis; "Different Worlds"; Essential Tension; Extraordinary Science; Gestalt Switch; Incommensurability; Meaning Change; Normal Science; Paradigm; Puzzle Solving; Revolutions in Science; Textbook Science; World Changes
Total	**30**	**14**

Revolutions (1962/1996), that is, thirty out of eighty index entries or 37.5%, compared to theoretical discussion of Kuhn's image of science (fourteen out of eighty index entries or 17.5%). Indeed, Newton alone is mentioned 137 times in *The Structure of Scientific Revolutions* (1962/1996), which is significantly more than "incommensurability," which is mentioned 12 times (including "incommensurable"), and more than "scientific revolution," which is mentioned 128 times. Nowadays, historians and philosophers of science would consider the "Great Men" historiography no less Whiggish than the positivist historiography of genius that Kuhn fiercely criticized (McEvoy 2010, 23–52). For, when we look at the history of science, not merely to find anecdotes, but to understand scientific change, as Kuhn himself urged us to do, we see a gradual change over time with multiple actors involved and alternative theories on offer along the way (Mizrahi 2015d). In the case of Newton, for example, he did not bring about the demise of Aristotelian physics all by himself overnight. In fact, by Newton's time, Aristotelian physics was already dying. The new game in town, with which Newton engaged, was Cartesian physics (Mizrahi 2015a, 367), which was itself developed by the likes of Henricus Regius and Jacques Rohault.

2. DEFENDING THE KUHNIAN IMAGE OF SCIENCE

Part II of this volume consists of two attempts to defend the Kuhnian image of science. In chapter 5, "The Kuhnian Straw Man," Vasso Kindi argues that critics of Kuhn's image of science often attack a straw man. Even though Kuhn is credited with initiating the so-called historical turn in philosophy of science (Wray 2011, 87), Kindi argues that the Kuhnian image of science is not meant to be descriptive, which is why it is not supposed to be supported by historical evidence. Instead, Kindi argues, Kuhn's model is derived from first principles. In particular, on the Kuhnian image of science, Kindi writes, "revolutions presuppose the existence of normal science, which is necessary in order to provide the background of normalcy against which anomalies are to be detected and recognized for what they are—deviations from normalcy" (p. 100).

If Kindi is right about this, then the question is why think that there are such things as "scientific revolutions" and "normal science." In other words, according to Kuhn (2000, 126):

> The existence of normal science is a corollary of the existence of revolutions [. . .]. If it did not exist (or if it were nonessential, dispensable for science), then revolutions would be in jeopardy also.

An argument from first principles for the Kuhnian image of science, then, would run along the following lines:

(1) If there are scientific revolutions, then there is normal science.
(2) There are scientific revolutions.

Therefore,

(3) There is normal science.

Of course, "one philosopher's *modus ponens* is another philosopher's *modus tollens*" (Putnam 1994, 280), so critics of the Kuhnian image of science would question the distinction between "normal science" and "revolutionary science," as they have done (see, e.g., Toulmin 1970). In that case, they could deny the consequent of (1), that is, that there is normal science, and derive the negation of the antecedent of (1). Indeed, as I have mentioned above, there are empirical reasons to think that the picture of the history of science as a graveyard of theories that were discarded after scientific revolutions is no more than a philosopher's myth (Mizrahi 2013a; 2015d; 2016).

In chapter 6, "Kuhn, Pedagogy, and Practice: A Local Reading of *Structure*," Lydia Patton argues that a "paradigm" is supposed to give the "rules of the game" (p. 115). For this reason, Patton urges us to think of the Kuhnian image of science, not in semantic terms having to do with theories, but in practical terms having to do with scientific practice or know-how. On Patton's reading, unlike Kindi's, the Kuhnian image of science is not derived from first principle but rather inferred as the best explanation for the behavior of scientists and those who study scientific practice (Patton 2015; cf. Mizrahi 2015c).

From this practical point of view, then, a "paradigm" is a "textbook" that "makes scientific research and progress possible" (p. 115). If Patton is right about this, then the question becomes how science could have started in the first place. For if scientists "working within a paradigm gain access to phenomena only from within a given paradigm" (p. 119), then they could not have gained access to phenomena (e.g., celestial phenomena) without having a paradigm (e.g., geocentrism) in the first place, which means that there must have been a paradigm prior to the one from within which they can now work. But then there must have been a prior paradigm before that, and another before that, and so on, *ad infinitum*. One might try to avoid this infinite regress by positing the existence of preparadigms or protoscience, as Patton suggests (p. 127, endnote 23). However, if preparadigms are supposed to make protoscience possible, just as paradigms are supposed to make science possible (on Patton's interpretation of the Kuhnian image of science), then they would seem to raise the same sort of problem.

In my own work (Mizrahi 2013b), I have studied aspects of scientific practice but did not find much support for the Kuhnian image of science. In particular, I have looked at the "rules of the game" for rewarding scientific work with the Nobel Prize in order to test the following conceptions of scientific progress (Bird 2008, 279):

> *The epistemic conception of progress*: An episode constitutes scientific progress precisely when it shows the accumulation of scientific knowledge.
> *The semantic conception of progress*: An episode constitutes scientific progress precisely when it either (a) shows the accumulation of true scientific belief, or (b) shows increasing approximation to true scientific belief.
> *The functional-internalist conception of progress*: An episode shows scientific progress precisely when it achieves a specific goal of science, where that goal is such that scientists can determine its achievement at that time (e.g., solving scientific puzzles [as in Kuhn's image of science]).

While searching through Nobel Lectures in Physics, Chemistry, Physiology or Medicine, and Economic Sciences (nobelprize.org) for the terms "truth," "knowledge," and "puzzle solution," I have found that scientists talk about progress in terms of knowledge more than they talk about it in terms of either truth or puzzle solutions (Mizrahi 2013b). For instance, if we look at all the Nobel Lectures in Physics (Kajita 2015 and McDonald 2015), Chemistry (Lindahl 2015, Modrich 2015, and Sancar 2015), Physiology or Medicine (Campbell 2015, Ōmura 2015, and Tu 2015), and Economic Sciences (Deaton 2015) from 2015, then, using Voyant Tools (voyant-tools.org), we can

Table 1.3. Number of occurrences of the terms "knowledge," "truth," and "puzzle solution" in Nobel Lectures in Physics, Chemistry, Physiology or Medicine and Economic Sciences from 2015 (Source: nobelprize.org)

Conception of scientific progress	Term	Count
Epistemic	know*	57
Semantic	tru*	20
Functional-internalist	puzzle*	8

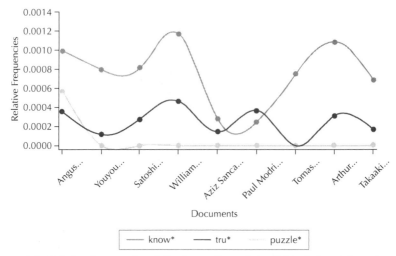

Figure 1.4. Relative frequencies of the terms "knowledge," "truth," and "puzzle solution" in Nobel Lectures in Physics, Chemistry, Physiology or Medicine and Economic Sciences from 2015 (Source: nobelprize.org).

see that the term "knowledge" is used more frequently than either "truth" or "puzzle solution" (see table 1.3 and figure 1.4).

As we can see from table 1.3 and figure 1.4, when they talk about scientific advancements that are worthy of the Nobel Prize in their Nobel Lectures, scientists use the term "knowledge" more frequently than the terms "truth" and "puzzle solution." For context, here is an example of "knowledge" talk in one of the Nobel Lectures from 2015 (Ōmura 2015, 258):

> As *science advances* and our *knowledge improves*, it is clear to me that the elucidation of suitable targets for medicines, and our expectations for finding

remedies to treat both *known* and as-yet *unknown* diseases and conditions, will not only *improve* but also *accelerate* (emphasis added).

Of course, the fact that scientists talk about scientific progress mostly in terms of knowledge does not necessarily imply that scientific progress consists in the accumulation of scientific knowledge. Nevertheless, if we take seriously the proposal to study carefully scientific practice, as Patton recommends, then the fact that scientists talk about progress mostly in epistemic ("knowledge") terms should (at the very least) inform our accounts of scientific progress. For this reason, these results count as *some* empirical evidence for the epistemic ("knowledge") conception of progress rather than the semantic ("truth") or the functional-internalist ("puzzle solution") conceptions of scientific progress.

3. REVISING THE KUHNIAN IMAGE OF SCIENCE

Part III of this volume consists of two chapters that seek to revise the Kuhnian image of science. In chapter 7, "Redefining Revolutions," Andrew Aberdein considers whether the Kuhnian image of science, suitably revised, can be applied to theory change in mathematics. His revision of the Kuhnian model consists in admitting that there are what he calls "glorious revolutions," that is, theoretical revolutions in mathematics that exhibit conceptual continuity and preservation of lexical taxonomies, and then restricting his account of theory change in mathematics to what he calls "inglorious revolutions," that is, theoretical revolutions in mathematics that exhibit replacement of lexical taxonomies, and "paraglorious revolutions," that is, theoretical revolutions in mathematics that exhibit conceptual continuity and preservation of key theoretical terms, as well as addition of new theoretical terms.

Aberdein then offers four cases studies: (a) the discovery of incommensurability (from rational to real numbers); (b) the rejection of the World Without End hypothesis in favor of the Doomsday hypothesis; (c) Mochizuki's claim to have proved the *abc* conjecture using his IUTeich theory; (d) the shift from elementary to advanced mathematics. Aberdein takes (a) to be a case of "at least paraglorious and perhaps also inglorious" revolution in mathematics (p. 149). He takes (b) to be a case of an inglorious revolution. He takes (c) to be a case of a paraglorious revolution, and (d) as initiating "numerous conceptual innovations, each of which might be seen as initiating a paraglorious revolution, and, when taken collectively, might represent a sorites-like inglorious revolution" (p. 149).

The reader will undoubtedly notice that Aberdein qualifies his claims about these putative cases of revolutions in mathematics, using words like "perhaps," "might," and "seem," and for good reasons. There are doubts about whether these are genuine cases of inglorious or paraglorious revolutions in mathematics. For example, as Aberdein points out, (a) may be more accurately described as a meta-mathematical revolution about "worldviews" rather than a mathematical revolution per se. As far as (b) is concerned, even if it is a case of referential discontinuity, as Aberdein (p. 144) claims, it does not necessarily follow that the Doomsday hypothesis and the World Without End hypothesis are conceptually incompatible, for referential discontinuity does not entail conceptual incompatibility (Mizrahi 2015a, 367). The problem with (c) is that this "revolution" never happened, as Aberdein (p. 139) points out. As for (d), Aberdein's figure 7.1 seems to show that there is a great deal of overlap between classical, modern, and contemporary mathematics, and all of these are subsumed under "advanced" mathematics. For the sake of argument, however, let us grant that Aberdein's case studies are genuine cases of inglorious and/or paraglorious revolutions in mathematics. The question still remains whether these case studies are representative of mathematics as a whole. As far as the whole of mathematics is concerned, are these case studies the exception or the rule?

In chapter 8, "Revolution or Evolution in Science? A Role for the Incommensurability Thesis?" James A. Marcum argues that scientific change must be thought of in evolutionary, not revolutionary, terms. As Marcum puts it: "Rather than the upheaval of world-shattering revolutions in the wake of science's advancement, scientific progress should be viewed as the gradual proliferation of scientific specialties much akin to biological speciation" (p. 159). On this evolutionary image of science, incommensurability becomes the mechanism of specialization, i.e., the means by which new scientific specialities come into being. Unlike Kuhn, however, who thought about the evolution of science in terms of Darwinian gradualism, Marcum argues that the tempos and modes of scientific evolution are more varied than that. In particular, Marcum identifies three tempos (Bradytelic, Tachytelic, and Horotelic) and three modes (Phyletic, Quantal, and Speciation) of scientific evolution.

As we have seen, there are a few problems with the evolutionary model of scientific change. For instance, as Park argues in chapter 3, a consequence of using evolutionary change as a model for scientific change seems to be that any theory of scientific change—if it is a product of evolution, like species are—will very likely die out, that is, will be discarded, given that "only about one in a thousand species is still alive—a truly lousy survival record: 99.9% failure!" (Raup 1992, 3–4). Accordingly, if biological evolution is our model for scientific evolution, we would expect to find evidence of mass extinctions of scientific theories in the historical record of science, just as we find evidence of mass extinctions of species in the fossil record. When we look at the historical record of science, however, we find no empirical evidence for mass extinctions of scientific theories, laws, or theoretical posits (Mizrahi 2013a; 2015d; 2016).

Nevertheless, I think that there is a way to test empirically Marcum's evolutionary model of scientific change. If Marcum is right about modeling scientific change on evolutionary change, then we would expect to find that scientific change is irreversible. This is because evolutionary change is irreversible. Once a new species has come into being, there is no going back. That is to say, the new species will not be able to interbreed successfully with the ancestral species. So, if scientific change is like evolutionary change, then it follows that practitioners of a new specialty would not be able to cooperate successfully with practitioners of the ancestral speciality in order to produce viable new scientific fields.

4. ABANDONING THE KUHNIAN IMAGE OF SCIENCE

Part IV of this volume consists of two chapters that go beyond the Kuhnian image of science in trying to find adequate models of scientific change. In chapter 9, "The Biological Metaphors of Scientific Change," Barbara Gabriella Renzi and Giulio Napolitano argue that orthogenesis is a better metaphor for scientific change than Darwinian evolution by natural selection. Renzi and Napolitano put the reason why orthogenesis is a better model for scientific change than Darwinian evolution as follows:

> In natural selective models, there is no correlation between occurring variations and the pressures of the organism's environment or its needs. On the other hand, changes in scientific ideas, which emerge as scientists try to solve problems, are the results of arguments and debates, the answers to the specific needs of a scientist or group of scientists who have been seeking a solution to a problem. This process of scientific change is more similar to orthogenesis than to Darwinian evolution (p. 184)

On the Darwinian model of evolution by natural selection, biological variation is unguided, primarily because genetic mutation is supposed to be random. On the orthogenetic model, however, variation is guided by the internal constitution of the organism. This is why orthogenesis is a better metaphor for scientific change, Renzi and Napolitano argue, for it reflects the way in which scientific development is guided by existing scientific knowledge and practices.

Any discussion of evolution raises questions about the units and levels of selection. In the case of biological evolution, the question is whether natural selection operates at the level of genes, organisms, species, groups, or populations. Since the "basic logic of natural selection is the same whatever the 'entities' in question are" (Okasha 2006, 10), those entities, that is, the units of selection, can in principle be genes, organisms, groups, species, and populations. This is why Renzi and Napolitano are right when they say that, no matter what the unit of selection is, the characteristics of selection are supposed to be the same. Nevertheless, just as it is an important question whether the units of natural selection are genes, organisms, groups, species, or populations, it is also an important question whether the units of scientific selection are concepts, posits, hypotheses, principles, theories, paradigms, research programs, disciplinary matrices, styles of reasoning, or something else entirely.

In chapter 10, "Beyond Kuhn: Methodological Contextualism and Partial Paradigms," Darrell P. Rowbottom seeks to move beyond the Kuhnian image of science by embracing the fact that science is a lot messier and more complicated than the Kuhnian image of science allows. In particular, Rowbottom proposes that normal science and extraordinary science "coexist" (p. 199). This is because different scientists perform different functions, such as puzzle solving, critical, and imaginative functions all at the same time. Depending on the state of science and its practitioners at a given time, more effort may be devoted to normal puzzle solving or to imaginative exploration. Rowbottom labels this view "methodological contextualism." For Rowbottom, this view is more attractive than the Kuhnian image of science because it makes science less rigid and more dynamic. In that respect, most of the contributors to this volume seem to agree that the Kuhnian image of science fails to capture all the complexities of this social and epistemic enterprise we call "science."

REFERENCES

Bird, Alexander. 2008. "Scientific Progress as Accumulation of Knowledge: A Reply to Rowbottom." *Studies in History and Philosophy of Science* 39 (2): 279–81.

Bird, Alexander. 2013. "Thomas Kuhn." In *The Stanford Encyclopedia of Philosophy*, edited by E. N. Zalta, Fall 2013 Edition. Accessed June 22, 2017. *https://plato. stanford.edu/archives/fall2013/entries/thomas-kuhn/*.

Brown, Joshua D. K. 2015. "Chemical Atomism: A Case Study in Confirmation and Ontology." *Synthese* 192 (2): 453–85.

Campbell, William C. 2015. "Nobel Lecture: Ivermectin: A Reflection on Simplicity." *Nobelprize.org*. Nobel Media AB 2014. Accessed June 26, 2017. http://www. nobelprize.org/nobel_prizes/medicine/laureates/2015/campbell-lecture.html.

Cartwright, Nancy. 1988. "A Case Study in Realism: Why Econometrics Is Committed to Capacities." *PSA: Proceedings of the Biennial Meeting of the Philosophy of Science Association* 1988 (2): 190–97.

D'Agostino, Fred. 2010. *Naturalizing Epistemology: Thomas Kuhn and the Essential Tension*. New York: Palgrave Macmillan.

Deaton, Angus. 2015. "Prize Lecture: Measuring and Understanding Behavior, Welfare, and Poverty." *Nobelprize.org*. Nobel Media AB 2014. Accessed June 26, 2017. http://www.nobelprize.org/nobel_prizes/economic-sciences/laureates/2015/deaton-lecture.html.

Fahrbach, Ludwig. 2011. "How the Growth of Science Ends Theory Change." *Synthese* 180 (2): 139–55.

Fuller, Steve. 1988/2002. *Social Epistemology*. Second Edition. Bloomington: Indiana University Press.

Green, Elliott. 2016. "What Are the Most-Cited Publications in the Social Sciences (according to Google Scholar)." *LSE Impact Blog*, May 12, 2016. Accessed May 24, 2017. http://blogs.lse.ac.uk/impactofsocialsciences/2016/05/12/what-are-the-most-cited-publications-in-the-social-sciences-according-to-google-scholar/.

Hettema, Hinne, and Theo A. F. Kuipers. 1995. "Sommerfeld's Atombau: A Case Study in Potential Truth Approximation." In *Cognitive Patterns in Science and Common Sense: Groningen Studies in Philosophy of Science, Logic and Epistemology*, edited by T. A. F. Kuipers and A. R. Mackor, 273–97. Amsterdam: Rodopi.

Hoyningen-Huene, Paul. 1993. *Reconstructing Scientific Revolutions: Thomas S. Kuhn's Philosophy of Science*. Chicago: The University of Chicago Press.

Hoyningen-Huene, Paul. 2006. "Context of Discovery Versus Context of Justification and Thomas Kuhn." In *Revisiting Discovery and Justification*, edited by J. Schickore and F. Steinle, 119–31. Dordrecht: Springer.

Hoyningen-Huene, Paul, and Eric Oberheim. 2009. "Reference, Ontological Replacement and Neo-Kantianism: A Reply to Sankey." *Studies in History and Philosophy of Science* 40 (2): 203–9.

Iranzo, Valeriano. 2002. "Constructive Empiricism and Scientific Practice. A Case Study." *Theoria* 17 (2): 335–57.

Kajita, Takaaki. 2015. "Nobel Lecture: Discovery of Atmospheric Neutrino Oscillations." *Nobelprize.org*. Nobel Media AB 2014. Access June 25, 2017. http://www.nobelprize.org/nobel_prizes/physics/laureates/2015/kajita-lecture.html.

Kindi, Vasso. 2015. "The Role of Evidence in Judging Kuhn's Model: On the Mizrahi, Patton, Marcum Exchange." *Social Epistemology Review and Reply Collective* 4 (11): 25–33.

Kindi, Vasso, and Theodore Arabatzis. eds. 2012. *Kuhn's* The Structure of Scientific Revolutions *Revisited*. New York: Routledge.

Kuhn, Thomas S. 1962/1996. *The Structure of Scientific Revolutions*. Third Edition. Chicago: The University of Chicago Press.

Kuhn, Thomas S. 1970. "Logic of Discovery or Psychology of Research?" In *Criticism and the Growth of Knowledge*, edited by I. Lakatos and A. Musgrave, 1–23. New York: Cambridge University Press.

Kuhn, Thomas S. 2000. "What Are Scientific Revolutions?" In *The Road since Structure*, edited by J. Conant and J. Haugeland, 13–32. Chicago: University of Chicago Press.

Kuhn, Thomas S. 2000. "Possible Worlds in History of Science." In *The Road since Structure*, edited by J. Conant and J. Haugeland, 58–89. Chicago: University of Chicago Press.

Kuhn, Thomas S. 2000. "Reflections on My Critics." In *The Road since Structure*, edited by J. Conant and J. Haugeland, 123–75. Chicago: University of Chicago Press.

Ladyman, James. 2011. "Structural Realism versus Standard Scientific Realism: The Case of Phlogiston and Dephlogisticated Air." *Synthese* 180 (2): 87–101.

Levine, Alex, and Mordechai Feingold. eds. 2010. "Special Issue: New Perspectives on Thomas Kuhn." *Perspectives on Science* 18 (3): 279–377.

Lindahl, Tomas. 2015. "Nobel Lecture: The Intrinsic Fragility of DNA." *Nobelprize. org*. Nobel Media AB 2014. June 26, 2017. http://www.nobelprize.org/nobel_prizes/chemistry/laureates/2015/lindahl-lecture.html.

Longino, Helen. 2016. "The Social Dimensions of Scientific Knowledge." In *The Stanford Encyclopedia of Philosophy*, edited by E. N. Zalta, Spring 2016 Edition. *https://plato.stanford.edu/archives/spr2016/entries/scientific-knowledge-social*.

Marcum, James A. 2015. "What's the Support for Kuhn's Incommensurability Thesis? A Response to Mizrahi and Patton." *Social Epistemology Review and Reply Collective* 4 (9): 51–62.

McDonald, Arthur B. 2015. "Nobel Lecture: The Sudbury Neutrino Observatory: Observation of Flavor Change for Solar Neutrinos." *Nobelprize.org*. Nobel Media AB 2014. Access June 26, 2017. http://www.nobelprize.org/nobel_prizes/physics/laureates/2015/mcdonald-lecture.html.

McEvoy, John G. 2010. *The Historiography of the Chemical Revolution*. New York: Taylor & Francis.

Mizrahi, Moti. 2013a. "The Pessimistic Induction: A Bad Argument Gone Too Far." *Synthese* 190 (15): 3209–26.

Mizrahi, Moti. 2013b. "What Is Scientific Progress? Lessons from Scientific Practice." *Journal for General Philosophy of Science* 44 (2): 375–90.

Mizrahi, Moti. 2015a. "Kuhn's Incommensurability Thesis: What's the Argument?" *Social Epistemology* 29 (4): 361–78.

Mizrahi, Moti. 2015b. "A Reply to James Marcum's 'What's the Support for Kuhn's Incommensurability Thesis?.'" *Social Epistemology Review and Reply Collective* 4 (11): 21–24.

Mizrahi, Moti. 2015c. "A Reply to Patton's 'Incommensurability and the Bonfire of the Meta-Theories.'" *Social Epistemology Review and Reply Collective* 4 (10): 51–53.

Mizrahi, Moti. 2015d. "Historical Inductions: New Cherries, Same Old Cherry-picking." *International Studies in the Philosophy of Science* 29 (2): 129–48.

Mizrahi, Moti. 2016. "The History of Science as a Graveyard of Theories: A Philosophers' Myth?" *International Studies in the Philosophy of Science* 30 (3): 263–78.

Modrich, Paul. 2015. "Nobel Lecture: Mechanisms in E. Coli and Human Mismatch Repair." *Nobelprize.org*. Nobel Media AB 2014. Access June 26, 2017. http://www.nobelprize.org/nobel_prizes/chemistry/laureates/2015/modrich-lecture.html.

Morris, Errol. 2011. "The Ashtray: The Ultimatum (Part 1)." *The New York Times*, March 6, 2011. *https://nyti.ms/2oNQmlH*.

Naughton, John. 2012. "Thomas Kuhn: The Man Who Changed the Way the World Looked at Science." *The Guardian*, August 18. https://www.theguardian.com/science/2012/aug/19/thomas-kuhn-structure-scientific-revolutions.

Oberheim, Eric. 2005. "On the Historical Origins of the Contemporary Notion of Incommensurability: Paul Feyerabend's Assault on Conceptual Conservatism." *Studies in the History and Philosophy of Science* 36 (2): 363–90.

Oberheim, Eric, and Paul Hoyningen-Huene. 2016. "The Incommensurability of Scientific Theories." In *The Stanford Encyclopedia of Philosophy*, edited by E. N. Zalta, Winter 2016 Edition. *https://plato.stanford.edu/archives/win2016/entries/incommensurability*.

Okasha, Samir. 2006. *Evolution and the Levels of Selection*. Oxford: Clarendon Press.

Ōmura, Satoshi. 2015. "Nobel Lecture: A Splendid Gift from the Earth: The Origins and Impact of Avermectin." *Nobelprize.org*. Nobel Media AB 2014. Access June 26, 2017. http://www.nobelprize.org/nobel_prizes/medicine/laureates/2015/omura-lecture.html.

Onkware, Kennedy. 2010. *Thomas Kuhn and Scientific Progression: Investigation on Kuhn's Account of How Science Progresses*. Staarbrücken: Lambert Academic Publishing.

Park, Seungbae. 2011. "A Confutation of the Pessimistic Induction." *Journal for General Philosophy of Science* 42 (1): 75–84.

Pashby, Thomas. 2012. "Dirac's Prediction of the Positron: A Case Study for the Current Realism Debate." *Perspectives on Science* 20 (4): 440–75.

Patton, Lydia. 2015. "Incommensurability and the Bonfire of the Meta-theories: Response to Mizrahi." *Social Epistemology Review and Reply Collective* 4 (7): 51–58.

Porter, Alan L., and Ismael Rafols. 2009. "Is Science Becoming More Interdisciplinary? Measuring and Mapping Six Research Fields Over Time." *Scientometrics* 81 (3): 719–45.

Preston, John. 2008. *Kuhn's* The Structure of Scientific Revolutions: *A Reader's Guide*. London: Continuum.

Putnam, Hilary. 1994. *Words and Life*, edited by James Conant. Cambridge, MA: Harvard University Press.

Raup, David M. 1992. *Extinction: Bad Genes or Bad Luck?* New York: W. W. Norton.

Sancar, Aziz. 2015. "Nobel Lecture: Mechanisms of DNA Repair by Photolyase and Excision Nuclease." *Nobelprize.org*. Nobel Media AB 2014. Access June 26, 2017. http://www.nobelprize.org/nobel_prizes/chemistry/laureates/2015/sancar-lecture.html.

Sankey, Howard. 1997. "Taxonomic Incommensurability." In *Rationality, Relativism and Incommensurability*, edited by H. Sankey, 66–80. London: Ashgate.

Scheffler, Israel. 1967. *Science and Subjectivity*. Indianapolis: Bobbs-Merrill.

Soler, Lena, Sjoerd Zwart, Michael Lynch, and Vincent Israel-Jost. 2014. "Introduction." In *Science after the Practice Turn in the Philosophy, History, and Social*

Studies of Science, edited by L. Soler, S. Zwart, M. Lynch, and V. Israel-Jost, 1–43. New York: Routledge.

Sterelny, Kim, and Paul E. Griffiths. 1999. *Sex and Death: An Introduction to Philosophy of Biology*. Chicago: The University of Chicago Press.

Torres, Juan M., ed. 2010. *On Kuhn's Philosophy and Its Legacy*. Lisbon: Faculdade de Ciêcias da Universidade de Lisboa.

Toulmin, Stephen E. 1970. "Does the Distinction Between Normal and Revolutionary Science Hold Water?" In *Criticism and the Growth of Knowledge*, edited by I. Lakatos and A. Musgrave, 39–47. New York: Cambridge University Press.

Tu, Youyou. 2015. "Nobel Lecture: Artemisinin—A Gift from Traditional Chinese Medicine to the World." *Nobelprize.org*. Nobel Media AB 2014. Accessed June 26, 2017. http://www.nobelprize.org/nobel_prizes/medicine/laureates/2015/tu-lecture.html.

Wray, Brad K. 2011. *Kuhn's Evolutionary Social Epistemology*. New York: Cambridge University Press.

Part I

QUESTIONING THE KUHNIAN IMAGE OF SCIENCE

Chapter 1

Kuhn's Incommensurability Thesis

What's the Argument?

Moti Mizrahi

1. INTRODUCTION

Since Thomas Kuhn first introduced the incommensurability thesis in 1962, it has undergone several transformations, even in Kuhn's own writings (Hoyningen-Huene 1993, 206–22).[1] Basically, the thesis has been understood in at least two different ways:

> *Taxonomic Incommensurability (TI)*: Periods of scientific change (in particular, revolutionary change) that exhibit TI are scientific developments in which existing concepts are replaced with new concepts that are *incompatible* with the older concepts. The new concepts are incompatible with the old concepts in the following sense: two competing scientific theories are conceptually incompatible (or incommensurable) just in case they do not share the same "lexical taxonomy." A lexical taxonomy contains the structures and vocabulary that are used to state a theory (see Kuhn 2000, 14–15, 63, 92–97, 229–33, 238–39, and 242–44. Cf. Sankey 1997).[2]
>
> *Methodological Incommensurability (MI)*: There are no objective criteria of theory evaluation. The familiar criteria of evaluation, such as simplicity and fruitfulness, are not a fixed set of rules. Rather, they vary with the currently dominant paradigm (see Kuhn 1962, 94; 1970, 200; 1977, 322, and 331. Cf. Sankey and Hoyningen-Huene 2001: xiii).[3]

Kuhn's incommensurability thesis has generated an enormous literature.[4] In this paper, I wish to examine the evidence that is supposed to support Kuhn's incommensurability thesis. This is an important task, I think, for the following reasons. First, some authors use the language of "discovery" when writing about Kuhn's incommensurability thesis. According to Oberheim and Hoyningen-Huene (2013), for instance, Kuhn "*discovered* incommensurability in the mid to late 1940s" (emphasis added). This "discovery" was then followed by a change in career paths. Using the "discovery" language in talking about Kuhn's incommensurability thesis gives the impression that incommensurability is a fact about scientific change (revolutionary change, in particular). But what if one does not think that incommensurability is a fact about scientific change? Are there compelling epistemic reasons to think that the incommensurability thesis is indeed true or probable? In other words, are there good *arguments* for the incommensurability thesis?

Second, and perhaps as a consequence of the first reason, it seems that the literature has focused more on exploring the implications of Kuhn's incommensurability thesis than on evaluating the *arguments* for the thesis itself.[5] This is not to say that no one has disputed Kuhn's incommensurability thesis (see, e.g., Scheffler 1982). Arguing that a thesis is false, however, is not the same as arguing that there is neither valid deductive nor strong inductive support for a thesis.

In what follows, I focus on (TI) rather than (MI). Here is how I plan to proceed. In section 2, I argue that there is no valid deductive argument[6] for (TI), since from the fact that the reference of the same kind terms changes or discontinues from one theoretical framework to another, it does not necessarily follow that these two theoretical frameworks are taxonomically incommensurable. In section 3, I argue that there is no strong inductive argument[7] for (TI) because there are rebutting defeaters against (TI). That is, there are episodes from the history of science that do not exhibit discontinuity and replacement, as (TI) predicts but rather continuity and supplementation. If this is correct, then there are no compelling arguments in support of (TI). In section 4, I discuss and reply to an objection made by an anonymous referee of the journal *Social Epistemology*, where this chapter was originally published.

2. WHY THERE IS NO VALID DEDUCTIVE SUPPORT FOR (TI)

Is there deductively valid support for (TI)? That is, is there a valid deductive inference from reference change to incompatibility of conceptual content, and thus to (TI)? By appealing to the distinction between sense and reference, Scheffler (1982, 59–60) has already shown that variation of sense does not entail incompatibility of conceptual content.[8] In that case, perhaps reference change is supposed to entail incompatibility of conceptual content, and hence taxonomic incommensurability. That is, "Distinct taxonomic structures (ones with different subsumption and exclusion relations) are inevitably incommensurable, *because those very differences result in terms with fundamentally disparate meanings*" (Conant and Haugeland 2000, 5; emphasis added). Indeed, in his later works, Kuhn talks about reference change as "redistribution" of members among "taxonomic categories" (Kuhn 2000, 30). Such "*referential changes*," according to Kuhn (2000, 15), show that "scientific development cannot be quite cumulative," that "one cannot get from the old to the new simply by an addition to what was already known," and that one cannot "*describe the new in the vocabulary of the old* or vice versa" (emphasis added). Furthermore, as Sankey (2009, 197) points out:

> Analysis of the reasoning employed by Kuhn [. . .] when [he] argue[s] for [. . .] reference change reveals that [he] assume[s] that reference is determined by description.

Accordingly, I take it that a deductive argument for (TI) would run roughly like this:

(TI1) If competing theories were taxonomically commensurable, then terms would still refer to the same things in new theories (e.g., "mass" in Newtonian mechanics and "mass" in relativistic mechanics would have the same referent).

(TI2) Terms do not refer to the same things in new theories (e.g., "mass" in Newtonian mechanics and "mass" in relativistic mechanics do not have the same referent, and some terms, such as "phlogiston," are eliminated outright).

Therefore:

(TI3) Competing theories are taxonomically incommensurable.[9]

In addition to "mass," another example that is supposed to illustrate how reference change is evidence for incompatibility of conceptual content is the following:

> Whereas Ptolemaic astronomers used the term "planet" to denote wandering stars, that is, those "stars" that are not fixed stars, Copernicus used the term "planet" to denote a celestial body that orbits the sun (Wray 2011, 25).

This variation in the referents of the term "planet," then, is supposed to show that the Ptolemaic model and the Copernican model are conceptually incompatible.

For this argument to be deductively valid, however, (TI3) must follow necessarily from (TI1) and (TI2). The crucial premise is (TI1). The claim here seems to be that T_1 and T_2 are taxonomically commensurable if and only if T_1 and T_2 share the same lexical taxonomy. Sharing the same lexical taxonomy, as we have seen, means that the kind terms that are used to state T_1 and T_2 refer to the same things. That is to say, for (TI1) to be true, the following conditionals would have to hold:

(C1) If kind terms t refer to X in T_1 and to X in T_2, then T_1 and T_2 are taxonomically commensurable.

(C2) If kind terms t refer to X in T_1 and to Y in T_2, then T_1 and T_2 are taxonomically incommensurable.

As Hoyningen-Huene (1993, 99) explains, to ensure that they refer to the same things with the same kind terms, scientists need only share vocabularies that incorporate the same taxonomies. If two competing theories do not share the same lexical taxonomies, then those two theories are taxonomically incommensurable (Kuhn 2000, 63).

However, as a general claim about the conceptual incompatibility, and hence incommensurability, of theoretical frameworks,[10] I think that (C2) is false. To see why (C2) is false, consider the following:

(K1) In biological taxonomy, "kid" refers to a young goat, whereas in folk taxonomy, "kid" refers to a child.

Therefore:

(K2) Biological and folk taxonomies are taxonomically incommensurable.

Since (K2) seems false, given that "kid" refers to a young goat in the order Artiodactyla but not in the order Primate, and both orders belong to the class Mammalia, even as far as folk taxonomy is concerned, there must be something wrong with the inference from (K1) to (K2). In other words, if the inference from (K1) to (K2) is supposed to be deductively valid, but (K2) is false, and (K1) is true, then it follows that the inference must be invalid after all, since valid inferences with true premises cannot lead to false conclusions. If this is correct, then from the fact that the same kind terms refer to different things in different theoretical frameworks, it does not necessarily follow that these theoretical frameworks are taxonomically incommensurable.[11]

To put it another way, if reference change were conclusive evidence for incompatibility of conceptual content, and hence for taxonomic incommensurability, then we would have to conclude that folk taxonomy and biological taxonomy are conceptually incompatible, since in the former, "kid" refers to one thing, and in the latter, "kid" refers to something else. But folk taxonomy and biological taxonomy are not conceptually incompatible. Therefore, reference change is not conclusive evidence for incompatibility of conceptual content. That is, there is no valid deductive inference from reference change to incompatibility of conceptual content, which means that there cannot be a valid deductive argument for (TI) from reference change alone.

For Kuhn, revolutionary change consists in the replacement of a lexical taxonomy by another incompatible taxonomy (Wray 2011, 25). That is to say, revolutionary changes violate Kuhn's (2000, 92) no-overlap principle, according to which:

> no two kind terms, no two terms with the kind label, may overlap in their referents unless they are related as species to genus. There are no two dogs that are also cats, no gold rings that are also silver rings, and so on: that's what makes dogs, cats, silver, and gold each a kind.

Individuating kind terms, however, is a rather tricky business. Is "puppy" a kind term? If so, "puppy" and "dog" are two kind terms, which overlap in their referents even though they are not related as species to genus. Be that as it may, for present purposes, the important point is that there are no conclusive epistemic reasons to think that scientific change, revolutionary or otherwise, involves abandoning a lexical taxonomy in favor of another incompatible one, given that reference change alone is not conclusive evidence for incompatibility of conceptual content, as the "kid" argument shows.

Proponents of (TI) might insist that deductively valid support for (TI) comes from periods of scientific change in which kind terms are dropped entirely (e.g., "phlogiston") rather than change their meaning (e.g., "mass"). In other words, it is not reference change (i.e., when the referents of kind terms change) that entails incompatibility of conceptual content but rather reference discontinuity (i.e., when kind terms no longer refer because they were abandoned).[12] In that case, the claim expressed by (TI1) would be that T_1 and T_2 are taxonomically commensurable if and only if the same kind terms from T_1 are retained in T_2. That is to say, for (TI1) to be true, the following conditionals would have to be true:

(C3) If kind terms t are used in T_1 and retained in T_2, then T_1 and T_2 are taxonomically commensurable.

(C4) If kind terms t are used in T_1 but dropped in T_2, then T_1 and T_2 are taxonomically incommensurable.

However, as a general claim about the conceptual incompatibility, and hence incommensurability, of theoretical frameworks,[13] I think that (C4) is false. To see why (C4) is false, consider the following:

(M1) In desktop environment, WIMP refers to a user interface that includes windows, icons, menus, and pointer, whereas in tablet environment, WIMP has been dropped entirely.

Therefore:

(M2) Desktop environment and tablet environment are taxonomically incommensurable.

Since (M2) is false, given that the same theoretical principles of computing govern the operation of both desktops and tablets, there must be something wrong with the inference from (M1) to (M2). In other words, if the inference from (M1) to (M2) is supposed to be deductively valid, but (M2) is false, and (M1) is true, then it follows that the inference must be invalid after all, since valid inferences with true premises cannot lead to false conclusions. If this is correct, then from the fact that kind terms have been abandoned by a later theoretical framework, it does not necessarily follow that the earlier and the later theoretical frameworks are taxonomically incommensurable.

In fact, even in the case of "phlogiston," it is not obvious that the two competing theories in question are indeed taxonomically incommensurable. Applied to the case of "phlogiston," the alleged inference from reference discontinuity to taxonomic incommensurability would run roughly as follows:

(P1) In Stahl's theory of combustion, "phlogiston" refers to the substance that is responsible for combustion, whereas in Lavoisier's theory of combustion, "phlogiston" is dropped and "oxygen" refers to the gas required for combustion.

Therefore:

(P2) Stahl's theory of combustion and Lavoisier's theory of combustion are taxonomically incommensurable.

But (P2) does not necessarily follow from (P1). To see why, suppose that Stahl's theory of combustion is true, and so there is phlogiston, which is the substance that is responsible for combustion. But it could still be the case that there is oxygen as well. In fact, it might actually be a useful addition to Stahl's theory of combustion, since one of its problems was to explain why, when some metals were calcined, the resulting calx was heavier than the original metal. Some tried to explain this by saying that, in some metals, phlogiston has negative weight. Instead, they could have said that phlogiston is lost but oxygen is gained, which would have explained the heavier weight.[14]

If this is correct, then, contrary to the claim that Stahl's theory of combustion and Lavoisier's theory of combustion are taxonomically incommensurable, since old concepts, such as phlogiston, were replaced by new concepts, such as oxygen, which are incompatible with the old concepts, the new concepts can actually *supplement*, rather than *replace*, the old ones. If this is correct, then it is not the case that (TI3) follows necessarily from (TI1) and (TI2).

To sum up, then, the considerations put forth in section 2, if correct, show that, just as variation of sense does not entail incompatibility of conceptual content (Scheffler 1982, 59–60), reference change also does not entail incompatibility of conceptual content. This is so because the fact that the referents of kind terms change from one theoretical framework to another does not entail that the two theoretical frameworks are conceptually incompatible or taxonomically incommensurable. Even discontinuity of reference does not entail incompatibility of conceptual content, since from the fact that a kind term from one theoretical framework no longer refers in a successive theoretical framework, it does not necessarily follow that these theoretical frameworks are conceptually incompatible or taxonomically incommensurable.

Some might think that Kuhn gives a better argument for (TI) in his (2000, 13–32), where Kuhn presents three examples of revolutionary change: Aristotelian and Newtonian conceptions of motion, the contact theory and the chemical theory of the Voltaic cell, and derivations of the law of blackbody radiation. On closer inspection, however, I think it becomes clear that Kuhn simply gives another version of the arguments from reference change and reference discontinuity, albeit using three different examples. For, as Kuhn himself puts it, the first two examples are "revolutions [that] were accompanied by changes in the way in which terms like 'motion' and 'cell' *attached to nature*" (emphasis added). As such, these are instances of reference change. The third example involves what Kuhn calls a "vocabulary change," for example, from "resonator" to "oscillator." As such, this is an instance of reference discontinuity.

Moreover, in the spirit of Friedman's (2011, 433) call "to *relativize* the Kantian conception of *a priori* scientific principles to a particular theory in *a given historical context*" (emphasis added) and "to *historicize* the notion of scientific objectivity" (original emphasis), one might question in what sense Aristotelian physics and Newtonian physics were "competing theories." After all, Newton did not seek to replace Aristotelian physics. According to Westfall (1983, 14):

> As far as men active in the study of nature were concerned, the word "overthrown" is not too strong. For them, Aristotelian philosophy was dead beyond resurrection.

Newton's target was not Aristotelian physics, but rather Cartesian physics, as the title of his *Principia* suggests (Brown 2005, 153–54). Likewise, insofar as Copernicus' heliocentric model had a competitor, it was not Ptolemy's geocentric model, but rather Tycho Brahe's hybrid model (Brown 2005, 156). When we take these historical facts into account, the transition from Aristotelian physics to Cartesian physics and then to Newtonian physics and the transition from the geocentric model to Brahe's hybrid model and then to the heliocentric model do not seem all that "revolutionary" (in the Kuhnian sense).

Accordingly, the argument of section 2 can be summed up as follows:

(1) Reference change is conclusive evidence for (TI) only if reference change entails incompatibility of conceptual content.
(2) Reference change does not entail incompatibility of conceptual content.

Therefore:

(3) It is not the case that reference change is conclusive evidence for (TI).

A similar argument applies to discontinuity of reference as well. That is:

(1) Reference discontinuity is conclusive evidence for (TI) only if reference entails incompatibility of conceptual content.
(2) Reference discontinuity does not entail incompatibility of conceptual content.

Therefore:

(3) It is not the case that reference discontinuity is conclusive evidence for (TI).

These arguments, if sound, show that there is no deductively valid inference from either reference change or reference discontinuity to (TI).[15] Accordingly, if a compelling argument can be made for (TI), that argument cannot be a deductively valid one. That leaves the option of a strong inductive argument. In the next section, then, I wish to see whether or not a strong inductive argument can be made in support of (TI).

3. WHY THERE IS NO STRONG INDUCTIVE SUPPORT FOR (TI)

If the considerations put forth in section 2 are correct, then there can be no valid deductive argument for (TI) from reference change and/or discontinuity alone. The other option, then, is a non-deductive or inductively strong argument for (TI). Accordingly, assuming for the sake of argument that there are episodes of scientific change in which competing theories exhibit taxonomic incommensurability, as Kuhn argues, the question is whether one can reasonably draw a general conclusion about the nature of scientific change (more precisely, revolutionary change) from such episodes. So, I take it that an inductive argument for (TI) would run roughly like this:

(TI4) Some episodes from the history of science exhibit taxonomic incommensurability (e.g., the Newtonian-Relativistic Mechanics episode, the phlogiston-oxygen episode).

Therefore:

(TI3) Scientific change (specifically, revolutionary change) is characterized by taxonomic incommensurability. (In other words, competing theories are taxonomically incommensurable.)

Now, even if we grant that Newtonian mechanics and relativistic mechanics are taxonomically incommensurable, I think that it is a mistake to generalize from a few selected examples that competing theories in general are taxonomically incommensurable. In the interest of brevity, I will briefly discuss one episode of scientific change. This episode is sufficient to show that one can reach different conclusions about the alleged taxonomic incommensurability of competing theories depending on the examples one (cherry-) picks. It is important to note that the following episode is not supposed to be a counterexample against (TI). It is not meant to be a refutation of (TI). Rather, it shows that an inductive argument based on a few selected historical episodes of scientific change does not provide strong inductive support for (TI). Or, to put it another way, this episode—and others like it—counts as what Pollock (1987, 481–518) calls a *rebutting defeater*, that is, a prima facie reason to believe the negation of the original conclusion; in this case, the negation of (TI).

Anastomoses

Early Modern physiologists faced the following problems, which were an inheritance from Galen: (a) How does blood flow from the right ventricle to the left? (b) Does the venous artery contain blood or air? According to Galen, the veins and arteries are two distinct systems. The only point of communication is the permeable septum in the heart. The problem was that Galen's conviction that the septum is permeable is based more on speculation than empirical evidence. Apparently, Vesalius was aware of this problem. In his *De Humani corporis fabrica* (1543), he writes:

The septum of the ventricles, therefore, is [. . .] made out of the thickest substances of the heart and on both sides is plentifully supplied with small pits which occasion its presenting an uneven surface towards the ventricles. Of these pits not one (at least in so far as is perceptible to the senses) penetrates from the right ventricle to the left, so that we are greatly forced to wonder at the skill of the Artificer of all things by which the blood sweats through passages that are invisible to sight from the right ventricle to the left (O'Malley 1965).

And in the 1555 edition, Vesalius writes:

Nevertheless, howsoever conspicuous these pits may be, not one of them, in so far as is perceptible to the senses, penetrates through the ventricular septum from the right to the left ventricle. Indeed, I have never come upon even the most obscure passages by which the septum of the heart is traversed, albeit that these passages are recounted in detail by professors of anatomy seeing that they are utterly convinced that the blood is received into the left ventricle from the right. And so it is (how and why I will advise you more plainly elsewhere), that I am not a little in two minds about the office of the heart in this respect (O'Malley 1965).

For Galen, much of the blood in the left ventricle comes from the right ventricle through these pores or pits in the septum. As Vesalius points out, however, no such pores or pits could be detected in the septum.

Following Vesalius, other anatomists became increasingly dissatisfied with the Galenist account of such pits. To name one of them, Realdus Columbus describes the pulmonary route of the blood from the right ventricle to the left in *De re anatomica libri xv* (1559):

Between these ventricles there is placed the septum through which almost all authors think there is a way open from the right to the left ventricle; and according to them the blood is in transit rendered thin by the generation of the vital spirits in order that the passage may take place more easily. But these authors make a great mistake; for the blood is carried by the artery-like vein to the lungs and being there made thin is brought back thence together with air by the vein-like artery to the left ventricle of the heart (Hall 1962, 273).

Although Juan Valverde da Hamusco was the first to publish experimental work on the minor circulation of the blood, the work of Columbus was widely read (Hall 1962, 274).

By the time William Harvey began his work on the circulation of the blood, the problems, if not the rudiments of a solution, were already set for him by his predecessors. After all, in addition to Valverde and Columbus' work on the minor circulation, Fabricius' *De venarum ostiolis* (1603) was known to Harvey. As a student of Fabricius at the University of Padua, Harvey was likely influenced by Fabricius' work, even though Fabricius did not publish his work until 1603 (Debus 1954, 63–65). For Fabricius, these *ostiola* are little doors that obstruct the flow of blood, so that the veins are not ruptured.

Although they would have made a very useful addition to his theory, Harvey was careful in his speculations and modest in his claims about the unobserved "capillaries." As Elkana and Goodfield (1968, 62) show, "Harvey was aware that the circulation was completed *by some means or other*; but throughout his life he was in considerable doubt as to how this was actually accomplished" (original emphasis). Harvey's reluctance to posit the existence of unobserved capillaries is evident in *Anatomical Disquisitions on the Circulation of the Blood* (1648), in which he replies to objections made by the French physician, Jean Riolan the Younger. As Harvey writes: "neither our learned author himself [i.e., Riolan], nor Galen, nor any experience, has ever succeeded in making such anastomoses as he imagines, sensible to the eye" (Bowie 1889, 107).

Harvey continues:

> I can therefore boldly affirm, that there is neither any anastomosis of the vena portae with the cava, of the arteries with the veins, or of the capillary ramifications of the biliary ducts, which can be traced through the entire liver, with the veins (Bowie 1889, 108).

So, according to his own accounts of the discovery of the circulation of the blood, it seems that Harvey was looking for such anastomoses as Galen's theory required. In doing so, he was clearly working within the framework of the problems that were set for him by his predecessors. After Harvey's death, Marcello Malpighi, studying the lungs of frogs using a microscope, showed that veins and arteries are joined by anastomoses (Elkana & Goodfield 1968; Allchin 2005).

Contrary to (TI), then, the anastomoses episode exhibits *continuity* in terms of the problems practitioners worked on, and *supplementation* in terms of old concepts that were abandoned for a while (e.g., anastomoses) but then rediscovered and added to the new theory, rather than discontinuity and replacement, as (TI) predicts.

The idea that some episodes of scientific change exhibit continuity is not new. For example, according to Friedman, there is continuity across scientific change at the level of what he calls "relativized *a priori*" principles. As Friedman (2001, 31) writes:

> Relativity theory involves *a priori* constitutive principles as necessary presuppositions of its properly empirical claims, just as much as did Newtonian physics, but these principles have essentially changed in the transition from the latter theory to the former: whereas Euclidean geometry is indeed constitutive *a priori* in the context of Newtonian physics, for example, only infinitesimally Euclidean geometry—consistent with all possible values of the curvature—is constitutively *a priori* in the context of general relativity. What we end up with,

in this transition, is thus a relativized and dynamical conception of *a priori* mathematical-physical principles, which change and develop along with the development of the mathematical and physical sciences themselves.

For Friedman (2001, 63), then, even though the concepts that constitute the relativized *a priori* principles change from one theoretical framework to another, and thus give rise to worries about incommensurability, there is still continuity, given that "earlier constitutive frameworks are exhibited as limiting cases, holding approximately in certain precisely defined special conditions, of later ones," and that the "principles of later paradigms [. . .] evolve continuously, by a series of natural transformations, from those of earlier ones." Indeed, some have explicitly drawn parallels between Friedman's relativized *a priori* and structural realism (see, e.g., MacArthur 2008. Cf. Ivanova 2011).

If the anastomoses episode is representative of scientific change, however, then it suggests that there can also be *conceptual continuity* (i.e., continuity in terms of concepts, such as anastomoses), not merely *structural continuity* (i.e., continuity in terms of structures or principles, such as Euclidean geometry and the equivalence principle)[16], across theory change. Adequately supporting this claim, however, requires a detailed examination of a greater variety of episodes of scientific change, which is beyond the scope of this paper. Indeed, one might complain that both Kuhn's and Friedman's accounts of scientific change are not representative of scientific change in general, since the case histories that are supposed to support these accounts are drawn from physics alone, to the exclusion of the social sciences and the life sciences. Be that as it may, my present aim is simply to show that there is no strong inductive support for (TI).

Generalizing from the history of science

In light of the anastomoses episode, which shows continuity and supplementation between competing theories, rather than replacement and incompatibility, as (TI) predicts, it seems fair to say that an inductive argument for (TI), construed as an inductive generalization, is an instance of hasty generalization. Indeed, as I have suggested above, the selected episodes that are supposed to support (TI), such as the phlogiston-oxygen episode, may even be unrepresentative of the nature of scientific change, since they are drawn exclusively from physics, thereby neglecting the life and social sciences. Again, the anastomoses episode is not supposed to be a counterexample against (TI).[17] It is not meant to refute (TI). Rather, it shows that a few episodes from the history of science, even if they do exhibit taxonomic incommensurability, do not provide strong inductive support for (TI).[18]

Accordingly, the argument of section 3 can be summed up as follows:

(1) There is a strong inductive argument for (TI) only if there are no rebutting defeaters against (TI).
(2) There are rebutting defeaters against (TI).

Therefore:

(3) It is not the case that there is a strong inductive argument for (TI).[19]

Accordingly, since there can be no deductively valid argument for (TI), as I have argued in section 2, and there can be no inductively strong argument for (TI), as I have argued in section3, it follows that there can be no good arguments for (TI). In other words, since deductive and inductive support for (TI) exhaust all forms of epistemic support for a thesis, given the way I have characterized them, it follows that there are no compelling epistemic reasons to believe that (TI) is true or probable.

To this friends of (TI) might object that (TI) is not supposed to be a general thesis about the nature of scientific change (more specifically, revolutionary change).[20] Instead, friends of (TI) might argue, some episodes of scientific change exhibit taxonomic incommensurability, whereas others do not.[21] This move, however, raises the following problems. First, if (TI) is not a general thesis about the nature of scientific change, then what is its *explanatory* value? How does (TI) help us in terms of understanding the nature of scientific change? On most accounts of explanation, an *explanans* must have some degree of generality, either because it counts as a "law of nature" as in the Deductive-Nomological model (Hempel 1965, 248), or because it is statistically relevant as in the Statistical Relevance model (Salmon 1971), or because it provides a unified account of a range of different phenomena as in the Unificationist model (Kitcher 1989). But if (TI) has no degree of generality, then it is difficult to see what the explanatory value of (TI) is.

Second, if (TI) is not a general thesis about the nature of scientific change, then what is its *predictive* value? If (TI) amounts to the claim that some competing theories are taxonomically incommensurable, then it is difficult to see how (TI) can help us make predictive inferences, such as "(TI); T_1 and T_2 are competing theories; therefore, T_1 and T_2 are taxonomically incommensurable," which are the inferences we need to be able to make in order to understand scientific change. In other words, if (TI) amounts to the claim that some competing theories are taxonomically incommensurable, whereas others aren't, then (TI) is nothing more than an interesting observation about some episodes of scientific change, not a theory about scientific change. In that respect, it is no more useful, as an attempt to understand scientific change, than the observation that some episodes of scientific change exhibit fraud. From the fact that some alleged "scientific discoveries" turned out to be fraudulent, such as the Piltdown man incident (Russell 2003), we cannot learn anything general and/or useful about the nature of science. I doubt that friends of (TI) would want to relegate (TI) to the level of an interesting observation about science that has no explanatory and/or predictive value.

To be clear, my response to this objection goes beyond the exegetical question of whether Kuhn meant (TI) to be a general thesis about scientific change (more precisely, "revolutionary change") or not.[22] My argument is that, as far as theory change is concerned, if (TI) is the exception rather than the rule, that is, if it is not the case that (TI) holds true of *most* (though not *all*) episodes of theory change, particularly "revolutionary change," then (TI) has no explanatory and/or predictive value. If (TI) has no explanatory and/or predictive value, then there are no "pragmatic" grounds for accepting (TI).[23]

Now, if there are no compelling epistemic reasons to believe that (TI) is true or probable, given that there is neither valid deductive nor strong inductive support for (TI), as I have argued, and given that there are no "pragmatic" reasons to accept (TI), such as considerations of explanatory and predictive power, then it is difficult to see why anyone should believe that (TI) is true or accept (TI). Stripped of explanatory and predictive value, "it looks as if incommensurability is irrelevant to questions of theory choice" (Brown 2005, 158).

4. AN OBJECTION AND A REPLY

An anonymous referee of the journal *Social Epistemology*, where this chapter was originally published, objects that conceptual incompatibility is supposed to be evidence for variation of reference, not the other way around. If the referee is right, then the fact that kind terms from one theoretical framework are either replaced with incompatible kind terms or dropped altogether in a competing theoretical framework is supposed to be conclusive evidence that these kind terms refer to different things.

I think that there are several problems with the referee's objection. First, it seems to be at odds with the characterization of (TI) in section 1 and the textual evidence cited in sections 1 and 2. This is not a serious problem, however, given that Kuhn's incommensurability thesis is notoriously difficult to pin down. More importantly, even if the referee is right, the question "What is the argument for (TI)?" remains unanswered. The referee argues that conceptual incompatibility is conclusive evidence for variation of reference. If so, then conceptual incompatibility is taken for granted as a premise from which variation of reference is supposed to follow. But the question raised in this paper is what the argument for *conceptual incompatibility* or *taxonomic incommensurability* is. To say that conceptual incompatibility is just taken for granted as evidence for something else is to miss the question that this paper raises.

In that respect, note that conceptual incompatibility cannot simply be gleaned from case histories of theory change. After all, (TI) is a theoretical claim, partly because concepts are unobservable. Hence, conceptual incompatibly has to be inferred from something else. What is that something else from which conceptual incompatibility is supposed to follow? The referee does not say if by "variation of reference," s/he means *reference change* (i.e., when the same kind terms refer to different things in competing theoretical frameworks) or *reference discontinuity* (i.e., when kind terms are dropped from one theoretical framework to another).

On the one hand, if by "variation of reference," the referee means *reference discontinuity*, then that means that kind terms from one theoretical framework have been dropped by a competing theoretical framework. In that case, however, conceptual incompatibility cannot simply be gleaned from the two competing theories, since the kind terms that are supposed to be conceptually incompatible have been dropped. That is to say, how can we tell if kind term K in T_1 is conceptually incompatible with K in T_2 if there is no K in T_2?

On the other hand, if by "variation of reference," the referee means *reference change*, then that means that the same kind terms have different referents. But in that case, too, conceptual incompatibility cannot simply be gleaned from the two competing theories, since the two competing theories *appear* to be compatible given that they share the same kind terms. That is to say, how can we tell whether T_1 is conceptually incompatible with T_2 or not if they share the same kind term K? If one were tempted to reply as follows: "Yes, they share the same kind terms, but these kind terms refer to different things; hence, the two competing theories are conceptually incompatible," then my answer would be "Precisely." That is, reference change is supposed to be evidence for conceptual incompatibility.

It is also worth noting that even *reference change* cannot simply be gleaned from case histories of theory change. This is so because of the distinction between what words mean and what a speaker means in uttering these words (Grice 1969) or the distinction between *semantic reference* and *speaker's reference* (Kripke 1977). To illustrate, consider the case of "planet" discussed above (Wray 2011, 25). From looking at the way Ptolemaic astronomers used "planet" and comparing it to the way Copernican astronomers used "planet," we might think that "planet" refers to different things, and hence that the Ptolemaic model and the Copernican model are taxonomically incommensurable. But that does not *necessarily* follow, since, even if the speaker's reference of "planet" is different in these two competing models, the semantic reference may be the same. Arguably, that was the case with "planet." The reverse may also be the case. That is, even if the semantic reference of kind term K is different in two competing theories, the speaker's reference may be the same. If this is correct, then the semantic/speaker's reference distinction undercuts the alleged entailment from reference change to conceptual incompatibility just as the sense/reference distinction undercuts the alleged entailment from semantic change to conceptual incompatibility (Scheffler 1982, 59–60; Sankey 2009, 197).

5. CONCLUSION

In this chapter, I have argued that there is no epistemic support, in the form of either a deductively valid or an inductively strong argument, for Kuhn's thesis of taxonomic incommensurability (TI). Reference change and/or discontinuity do not provide conclusive evidence for (TI), since there is no valid deductive inference from reference change and/or discontinuity to incompatibility of conceptual content. Moreover, there can be no strong inductive argument for (TI), since there are episodes from the history of science that do not exhibit discontinuity and replacement, as (TI) predicts, but rather continuity and supplementation. These historical episodes do not *refute* (TI). Rather, they are rebutting defeaters against (TI).

If, in an attempt to save (TI), friends of (TI) retreat to the position that (TI) is not supposed to be a general thesis about the nature of scientific change (specifically, revolutionary change), then they have to face the consequence that there are no "pragmatic" grounds for accepting (TI), either, since (TI) would be stripped of any explanatory and predictive value. If this is correct, then there are no compelling epistemic reasons for (TI), which means that no one should believe that (TI) is true or probable, and there are no "pragmatic" reasons for (TI), which means that no one should accept (TI).

ACKNOWLEDGMENTS

An earlier version of this chapter was presented at the *Kuhn's* The Structure of Scientific Revolutions*: 50 Years On* conference, December 1, 2012, at the College of New Jersey. I would like to thank the organizers, Pierre Le Morvan and Rick Kamber, as well as the audience for helpful feedback. I am also grateful to two anonymous reviewers of *Social Epistemology* for helpful comments on earlier drafts.

NOTES

1 Another version of the incommensurability thesis was proposed by Paul Feyerabend around the same time. As Sankey (2009, 196) notes, however, "It is widely recognized that Kuhn and Feyerabend did not mean the same thing when they

originally spoke of the incommensurability of competing theories." For the purposes of this chapter, I focus on Kuhn's version of the incommensurability thesis.

2 On taxonomic incommensurability as a special kind of conceptual incompatibility in terms of lexical taxonomies, see Kuhn (2000, 14–15) and the introduction to Sankey and Hoyningen-Huene (2001).

3 For yet another notion of incommensurability, see Bird (2007, 21–39).

4 See, for example, Sankey (1994) and the chapters collected in Sankey and Hoyningen-Huene (2001).

5 For recent examples, see the "special issue of *Social Epistemology* [vol. 17, issue 2–3, 2003] devoted to critical comments on Fuller's study of the philosophy of Thomas S. Kuhn and its context" (Gattei 2003, 89).

6 By "valid deductive argument," I mean an argument whose premises, if true, necessitate the truth of the conclusion. An invalid argument, then, is an argument whose premises, even if true, do not necessitate the truth of the conclusion.

7 By "strong inductive argument," I mean an argument whose premises, if true, make the truth of the conclusion more likely but not guaranteed. A weak argument, then, is an argument whose premises, even if true, do not make the conclusion more probable.

8 On semantic incommensurability and variation of sense, see Sankey (2009, 197).

9 Cf. Stillwaggon Swan and Bruce (2011, 341–43).

10 (C2) seems false as a general claim about conceptual schemes as well. On paradigms as conceptual schemes, see Henderson (1994, 174). See also Putnam (1981, 114). Cf. Feyerabend (1987, 75–81).

11 Cf. Dupré's (1981) "lily" example. Note that although "lily" refers to one thing in plant taxonomy and to another thing in ordinary language, it also refers to both in these conceptual frameworks, which is why there is no conceptual incompatibility, and hence no taxonomic incommensurability, in this case (Dupré 1981, 74). Similarly, although "kid" refers to a young goat in animal taxonomy and to a child in ordinary language or folk taxonomy, it also refers to both in these conceptual frameworks, which is why there is no conceptual incompatibility, and hence no taxonomic incommensurability, in this case, either.

12 See Sankey (2009, 197) on semantic incommensurability and discontinuity of reference.

13 See endnote 10.

14 On the phlogiston-oxygen episode, see also Kitcher (1978) who argues that "dephlogisticated air" genuinely referred to samples of oxygen.

15 Cf. Sankey (2009). According to Sankey (2009, 198), "the threat of wholesale referential discontinuity between theories has been removed by rejecting the description theory of reference." That is, if "reference is independent of description," then "successive theories are not incommensurable due to discontinuity of reference" (Sankey 2009, 198).

16 These are Friedman's (2001, 71) examples.

17 Another episode from the history of science that runs contrary to (TI) is the scientific change from Wegener's theory of continental drift to plate tectonics.

18 Just as it would be hasty to conclude from a few episodes that incommensurability is typical of scientific change, it would also be hasty to conclude from a few episodes that incommensurability is atypical of scientific change.

19 Given what I mean by deductive and inductive arguments, any nondeductive form of inference is covered by "inductive" for the purposes of this paper, except abduction or Inference to the Best Explanation (IBE). I think it is safe to assume that incommensurability is not supposed to be the best explanation for scientific change, since (TI) is supposed to be *a mark of* scientific change, not *an explanation for* scientific change. Moreover, it is not clear what, if any, novel predictions (TI) makes. Cf. Mizrahi (2012).

20 By "general thesis," I mean a thesis that holds true of *most*—though not *all*—cases. So, if (TI) is not a general thesis, then it does not hold true even of *most* episodes of scientific change. This would be a problem, of course, if the argument for (TI) is supposed to be an inductive generalization.

21 See Kuhn (1962, 103). Cf. Bird (2012).

22 Though see Sankey (1993). Since (TI) is an integral part of Kuhn's theory of scientific revolutions, and that theory is supposed to be a general account of scientific change, it follows that (TI) is supposed to hold generally, too.

23 I use "pragmatic" and "acceptance" in the constructive empiricist's sense. On "acceptance," see van Fraassen (1980, 88). On the distinction between epistemic and nonepistemic (or pragmatic) values in theory choice, see van Fraassen (2007, 340).

REFERENCES

Allchin, D. 2005. William Harvey and capillaries. *The American Biology Teacher* 67(1): 56–59.

Bird, A. 2007. Incommensurability naturalized. In L. Soler, H. Sankey and P. Hoyningen-Huene (eds.), *Rethinking Scientific Change and Theory Comparison* (pp. 21–39). Berlin: Springer.

Bird, A. 2012. *The Structure of Scientific Revolutions* and its significance: An essay review of the fiftieth anniversary edition. *British Journal for the Philosophy of Science* 63, 859–83.

Bowie, A. (Ed.). 1889. *On the Motion of the Heart and the Blood in Animals by William Harvey M.D.* London: George Bell & Sons.

Brown, H. I. 2005. Incommensurability reconsidered. *Studies in History and Philosophy of Science* 36, 149–69.

Conant, J. and Haugeland, J. 2000. Editors' introduction. In J. Conant and J. Haugeland (eds.), *The Road since Structure* (pp. 1–10). Chicago: University of Chicago Press.

Debus, A. G. 1954. *Man and Nature in the Renaissance*. New York: Cambridge University Press.

Dupré, J. 1981. Natural kinds and biological taxa. *The Philosophical Review* 90(1): 66–90.

Elkana, Y. and Goodfield, J. 1968. Harvey and the problem of the "capillaries". *Isis* 59(1): 61–73.

Feyerabend, P. K. 1962. Explanation, reduction and empiricism. In H. Feigl and G. Maxwell (eds.), *Scientific Explanation, Space, and Time* (pp. 28–97). *Minnesota*

Studies in Philosophy of Science, vol. III. Minneapolis: University of Minnesota Press.

Feyerabend, P. K. 1987. Putnam on incommensurability. *British Journal for the Philosophy of Science* 38(1): 75–81.

Friedman, M. 2001. *Dynamics of Reason*. Chicago: University of Chicago Press.

Friedman, M. 2011. Extending the dynamics of reason. *Erkenntnis* 75, 431–44.

Gattei, S. 2003. Editor's introduction. *Social Epistemology* 17(2–3): 89–324.

Grice, P. 1969. Utterer's meaning and intentions. *The Philosophical Review* 78, 147–77.

Hall, M. B. 1962. *The Scientific Renaissance 1450–1630* (1994 ed.). New York: Dover Publications, Inc.

Hempel, C. 1965. *Aspects of Scientific Explanation and Other Essays in the Philosophy of Science*. New York: Free Press.

Henderson, D. K. 1994. Conceptual schemes after Davidson. In G. Preyer, F. Siebelt, and A. Ulfig (eds.), *Language, Mind, and Epistemology* (pp. 171–98). Dordrecht: Kluwer.

Hoyningen-Huene, P. 1993. *Reconstructing Scientific Revolutions: The Philosophy of Science of Thomas S. Kuhn*. Chicago: University of Chicago Press.

Hoyningen-Huene, P. 2008. Thomas Kuhn and the chemical revolution. *Foundations of Chemistry* 10, 101–15.

Ivanova, M. 2011. Friedman's relativised *a priori* and structural realism: In search of compatibility. *International Studies in the Philosophy of Science* 25(1): 23–37.

Kitcher, P. 1978. Theories, theorists and theoretical change. *The Philosophical Review* 87, 519–47.

Kitcher, P. 1989. Explanatory unification and the causal structure of the world. In P. Kitcher and W. Salmon (eds.), *Scientific Explanation* (pp. 410–505). Minneapolis: University of Minnesota Press.

Kripke, S. 1977. Speaker's reference and semantic reference. *Midwest Studies in Philosophy* 2, 255–76.

Kuhn, T. 1957. *The Copernican Revolution*. Cambridge, MA: Harvard University Press.

Kuhn, T. 1962. *The Structure of Scientific Revolutions*. Chicago: University of Chicago Press.

Kuhn, T. 1970. Postscript—1969. In *The Structure of Scientific Revolutions* (pp. 174–210), Chicago: University of Chicago Press.

Kuhn, T. 1977. *The Essential Tension*. Chicago: University of Chicago Press.

Kuhn, T. 2000. *The Road since Structure*. J. Conant and J. Haugeland (eds.), Chicago: University of Chicago Press.

McArthur, D. 2008. Theory change, structural realism, and the relativised *a priori*. *International Studies in the Philosophy of Science* 22, 5–20.

Mizrahi, M. 2012. Why the ultimate argument for scientific realism ultimately fails. *Studies in History and Philosophy of Science* 43, 132–38.

Oberheim, E. and Hoyningen-Huene, P. 2013. The incommensurability of scientific theories. In E. N. Zalta (ed.), *The Stanford Encyclopedia of Philosophy* (Spring 2013 Edition). http://plato.stanford.edu/archives/spr2013/entries/incommensurability/.

O'Malley, C. D. 1965. *Andreas Vesalius of Brussels 1514–1564*. Berkeley: University of California Press.

Pollock, J. L. 1987. Defeasible reasoning. *Cognitive Science* 11(4): 481–518.

Putnam, H. 1981. *Reason, Truth and History*. New York: Cambridge University Press.

Russell, M. 2003. *Piltdown Man: The Secret Life of Charles Dawson*. Stroud, UK: Tempus.

Salmon, W. 1971. Statistical explanation. In W. Salmon (ed.), *Statistical Explanation and Statistical Relevance* (pp. 29–87). Pittsburgh: University of Pittsburgh Press.

Sankey, H. 1993. Kuhn's changing concept of incommensurability. *British Journal for the Philosophy of Science* 44(4): 759–74.

Sankey, H. 1994. *The Incommensurability Thesis*. London: Ashgate.

Sankey, H. 1997. Taxonomic incommensurability. In H. Sankey (ed.), *Rationality, Relativism and Incommensurability* (pp. 66–80). London: Ashgate.

Sankey, H. and Hoyningen-Huene, P. 2001. Introduction. In *Incommensurability and Related Matters* (pp. vii–xxxiv), Dordrecht: Kluwer.

Sankey, H. 2009. Scientific realism and the semantic incommensurability thesis. *Studies in History and Philosophy of Science* 40(2): 196–202.

Scheffler, I. 1982. *Science and Subjectivity*. Indianapolis: Hackett Publishing Company.

Stillwaggon Swan, L. and Bruce, M. 2011. Kuhn's Incommensurability Arguments. In M. Bruce and S. Barbone (eds.), *Just the Arguments: 100 of the Most Important Arguments in Western Philosophy* (pp. 341–43). Malden, MA: Wiley—Blackwell.

Van Fraassen, B. 1980. *The Scientific Image*. New York: Oxford University Press.

Van Fraassen, B. 2007. From a view of science to a new empiricism. In B. Monton (ed.), *Images of Empiricism: Essays on Science and Stances, with a Reply from Bas C. van Fraassen*. New York: Oxford University Press.

Westfall, R. 1983. *Never at Rest: A Biography of Isaac Newton*. Cambridge: Cambridge University Press.

Wray, K. B. 2011. *Kuhn's Evolutionary Social Epistemology*. New York: Cambridge University Press.

Chapter 2

Modeling Scientific Development

Lessons from Thomas Kuhn

Alexandra Argamakova

1. THE KUHNIAN IMAGE OF SCIENCE

Soon after publication in 1962, ideas from *The Structure of Scientific Revolutions* came into intensive circulation and attracted lively attention. The book's concepts spread widely in public and terms such as "paradigm," "paradigm shift," and "scientific revolution" entered various disciplines and discourses, transforming Kuhn's original meaning. From 1962 to the present, *Structure* remains among the most influential, cited, and read philosophical works.[1] These facts of history might suggest that something similar to a Kuhnian paradigm emerged in the philosophy of science. However, the image of science presented in *Structure* never fully captured the intellectual market.

Significant and varied criticism quickly followed publication of the book.[2] *Structure* sparked vibrant debates regarding the theory of scientific revolutions. Depending on the particular analysis, Kuhn could be admired or criticized for the same things. Critics argued simultaneously against his naturalized methodology (Popper, Lakatos), and incomplete sociologization of his approach (Barnes, Bloor). Kuhn was blamed equally for a postmodernist disruption with tradition bringing relativism, subjectivism, and irrationalism to science studies (Shapere, Scheffler, Sokal) and persistent conservatism revealing in relapses to classical epistemological claims (Shapere, Barnes, Bloor); for commitment to the artificial schema of scientific growth (Feyerabend), and its particular specifications (Lakatos); for disregard of continuities in the history of science (Toulmin), and the conception of "normal" cumulative periods (Popper, Watkins, Feyerabend). Critics offered not only numerous counter-arguments but also full-blooded alternative theories of science: critical rationalism and falsificationism by Popper, the methodology of scientific research programs by Lakatos, the evolutionary model of science by Toulmin, epistemological anarchism by Feyerabend, historical epistemology by Wartofsky, and many more. The intellectual landscape shows great diversity in that period. Much recent literature lends both skepticism toward Kuhn and various attempts at updating his ideas.[3]

Thus, Kuhn's image of science can be characterized as influential rather than dominant. But what about this image that captured the attention of the intellectual public? On the one hand, *Structure* seems to be a primary source. On the other hand, its statements were reinterpreted and reconsidered in Kuhn's later writings, being selected and gathered in two collections of essays—*The Essential Tension* (1977) and *The Road since Structure* (2000). More likely, there exist various images of science belonging to different Thomas Kuhns at different stages of his work life and from different perspectives of interpretation, so the target for current analysis turns out to be less detectable.

Though much was said and written on the topic, a fresh assessment of Kuhn's legacy can show how far we all have advanced on *The Road since Structure* and where deadlock and promising pathways lie in the history and philosophy of science nowadays. Kuhn gave strong impulses to social studies of science[4] and revised basic principles of classical epistemology. Still, various lessons remain to be learned from his work. My analysis will focus on two issues. First, why the theory of scientific revolutions does not survive under the weight of counter-arguments even if its particular insights remain of interest. Three strategies for disruptive critiques will be presented in terms of both their general features and significant details. Second, a concern that arises from the first issue involves the validity of meta-theorizing and universal generalizations about scientific practice and history. This concern is often ignored in Kuhn studies, though it may reveal significant reasons to be dissatisfied with his philosophy of science.

2. THREE STRATEGIES FOR CRITIQUE

The theory of scientific revolutions pictures the development of science via two phases, cumulative and revolutionary. The cumulative phase is also called "normal science" and concretized through the idea of a paradigm; that is, an established conceptual framework and rules for conducting research. Research within a paradigm is not marked by novelty but is occupied with routine puzzle-solving activities. On the contrary, the revolutionary phase, or extraordinary science, is a period of radical creativity, paradigm change and the move to a new framework and methodology for a discipline. The move happens during the time of intense crisis, provoked by the appearance of anomalies. The means of old theories do not work for explaining anomalies and new theories come into existence to cope with a crisis. In addition, Kuhn includes a dose of sociological theory and stresses the significance of social factors regarding the behavior of scientific communities in the process of learning and changing paradigms.

Revolutionary episodes in the history of science are not consistent with a cumulative account of scientific practice. Old-fashioned cumulativism is one of Kuhn's main targets, and he displaces such approaches by formulating a thesis about the incommensurability of old and new paradigms. According to the incommensurability thesis, a scientific revolution is not just the rapid expansion of knowledge and appearance of new theories following a period of crisis. A scientific revolution comes together with certain shifts and losses on theoretical, methodological, and axiological levels in meanings, problems, statements, assumptions, principles, instruments, and the values of scientists. Kuhn avoids discussing theory-change in terms of truth. In this sense, the new paradigm is no better than the previous one. This equivalence does not mean Kuhn slips into relativism. More believable, as Ian Hacking (2012, xxxv) notes, Kuhn "did reject a simple 'correspondence theory' which says true statements correspond to facts about the world." The choice between competing theories is based on a more complex system of considerations than a comparison of truth-values; rather, it is determined by a mix of cognitive and social factors (Kuhn 1970b, 241).

The theory of scientific revolutions could not be less attractive for so many minds! At first glance, it is a clear, elegant, and testable theory developed by Kuhn with the instinct of a physicist looking for a simple "all-purpose" structure of how science changes in history (Hacking 2012, xi). However, one can find a great number of disruptive arguments against it. The three main strategies for critique are to disclose the *ambiguity*, *inaccuracy*, or *limitation* of Kuhn's model.

Ambiguity

The curious fact about language is that its conceptual ambiguity does not prevent its usage, even in scientific or philosophical theory. Language should not be more precise than we need for particular purposes. At the same time, for cognitive purposes, ambiguity is regarded as a serious defect, especially if it is unresolvable. This sort of conceptual ambiguity can be found for each central notion of *Structure*.

In a critical response to Kuhn, Masterman (1970, 59–89) explicates twenty-one senses of how the word "paradigm" is used in *Structure*. Kuhn objects to her with twenty-two senses. Since they do not contradict each other, as Kuhn argues in the Postscript to the second edition of the book (1970c, 174–210), they are summarized in four groups. Paradigm, renamed now as a disciplinary matrix, embraces

(1) symbolical generalizations and conceptual models;
(2) metaphysical assumptions;
(3) values of scientists directing research;
(4) exemplars; that is, "concrete problem solutions, the sorts of standard examples of solved problems which scientists encounter first in student laboratories, in the problems at the ends of chapters in science texts, and on examinations" (Kuhn 1970b, 272).

This list omits the methodological constituent of paradigm, which was emphasized before and may be just partly embedded in the idea of an exemplar. Afterward, Kuhn concentrates attention on developing the narrow understanding of paradigm as an exemplar and directly acknowledges that he "lost control of the word" (Kuhn 1970b, 272), which absorbed too many meanings.

Paradigm is something shared by a scientific community. Defining it as a disciplinary matrix, Kuhn states that it functions on the level of a discipline. The question is whether that means certain fields of science (e.g.,biology, physics, and chemistry) or branches within science (in physics, for example, mechanics, optics, and electrodynamics). Moreover, some fundamental theories belong to several disciplines at once. The theory of evolution develops within evolutionary biology, embryology, genetics, molecular biology, and biochemistry. The theory of artificial intelligence is essentially an interdisciplinary enterprise developing at the intersection of various branches of natural and human sciences. It seems impossible to apply the concept of disciplinary matrix to such cross-areas. Paradigmatic shift, occurring in evolutionary theory or the theory of artificial intelligence, must produce changes in several disciplines and disciplinary matrices, not just in a single one. Moreover, disciplines can include not only the same fundamental theories but also different approaches and instruments for research, as the earlier mentioned examples illustrate. This issue does not match very well with Kuhn's original thoughts on paradigm, disciplinary matrix, and the character of normal science.

Kuhn describes normal science as a routine puzzle-solving activity in which there is a lack of true creativity and novelty: "Normal science does not aim at novelties of fact or theory and, when successful, finds none" (Kuhn 1970b, 252). Scientific research as puzzle solving is a metaphorical expression and, in his later writings, Kuhn agreed with its vagueness as argued by critics (Kuhn 2000, 142). The application and development of general theories and laws form the essential part of scientific work but are not necessarily routine and puzzle solving in nature. When the electromagnetic theory of light was created on the basis of Maxwell's equations, it transformed the entire field of optics. Surely, one can always characterize it as a paradigmatic change rather than a creative development within the electromagnetic paradigm. But while Maxwell's electromagnetism constitutes a new paradigm, it is again unclear whether the electromagnetic theory of light, founded on Maxwell's equations, necessarily satisfies the criterion of a revolutionary shift according to Kuhn. Indeed, a list of precise criteria was never provided.

Not every theory-change is of a revolutionary nature. Revolutions presuppose the radical change in fundamental theories, methodologies, metaphysics, and scientific worldviews, accompanied by corresponding types of incommensurability. The problem is whether these changes must happen together or, maybe, it is not always the case. The theory of scientific revolutions does not give direct answers to these questions and it can be challenged by rebutting cases from history. As Barnes suggests:

> Few commentators would deny that the empirical characteristics of scientific revolutions are inadequately specified in Kuhn's *Structure*; and the lack of any development in this regard in the two subsequent decades greatly adds to the force of the criticism. Revolutions range from massive reconstructions extending over decades, to quickly accomplished cognitive and procedural reorientations such as are implied, for example, by the discovery of a new planet. They include changes in the common culture of the educated elite of the whole of Europe, and esoteric modifications in the accepted problem-solutions of small groups of highly specialised professionals. They are generally defined as retooling operations with important consequences for research practice, but they are occasionally treated more abstractly as changes in cosmology or worldview. One is bound to wonder why Kuhn has never sought to prune and discipline this initial diversity of sense (Barnes 1982, 56).

The mechanisms of paradigm-change, and the appearance of anomalies and crisis, are central themes of *Structure* that intersect with one another. Anomalies serve as the source of troubles and creative impulses for scientists leading to the new theories during crisis. An anomaly is what cannot be explained by the means of existing theories. It is an unknown fact, an unexpected result or data, contradicting common knowledge. The elimination of anomalies proceeds by the way of normalizing them within old theories or creating new explanations. Anomalies can give a stimulus for theory-change, but not every change of theory means a paradigm shift. According to Durkheim's holistic logic, borrowed by Kuhn, until the necessity of corrections in core theories is justified, anomalies can be eliminated through creation of auxiliary hypotheses and theories. Thus, three types of anomalies seem to appear:

(1) Anomalies normalized in old theories
(2) Anomalies normalized in new auxiliary theories
(3) Anomalies normalized in new fundamental theories (paradigms)

Kuhn never distinguished these three types, though they logically follow from his broad statements about anomalies in science. His intent was to describe an anomaly as type (3), but its specific meaning was not clarified and distinguished from (1) and (2).

There are other flaws, concerning conceptual ambiguity, which will be left untouched for now: the mix of revolutionary and evolutionary metaphors, the weak distinction between normative and descriptive components of the theory, the incommensurability of the incommensurability thesis, which has changed meanings many times (Sankey 1993), and so on. Yet, perhaps Kuhn should not be held responsible for clearing up these ambiguities. Pioneering works often include ambiguous vocabulary and raise good questions rather than provide answers. But further attempts at clarification did not achieve significant success or lead the followers to results significantly different from what Kuhn initially thought.

Inaccuracy

Vague vocabulary and misconceptions are not the same things, though inter-dependent. Does science really work as Kuhn describes? A further reason to be dissatisfied with Kuhn's theory is if it gives an inaccurate model of scientific development.

The initial inspiration for this book was Moti Mizrahi's (2015) article Kuhn's Incommensurability Thesis: What's the Argument? in *Social Epistemology*, and the subsequent discussion on the *Social Epistemology Review and Reply Collective* (social-epistemology.com), to which I now appeal. Incommensurability is one of Kuhn's most important ideas because it is inseparable from revolutions. Both ideas justify and support each other. In the end, these ideas allow one to draw a distinction between two phases of scientific development—normal (paradigmatic) and extraordinary (revolutionary) science. As Kuhn writes:

> Since the science which I call normal is precisely research within a framework, it can only be the opposite side of a coin the face of which is revolutions . . . . Frameworks must be lived with and explored before they can be broken (Kuhn 1970b, 242).

Besides, Kuhn says: "I have so far argued that, if there are revolutions, then there must be normal science. One may, however, legitimately ask whether either exists" (Kuhn 1970b, 249).

The following grounds for their existence are to be:

(1) Difference in the character of scientific research: cumulative normal science versus noncumulative revolutionary science, characterized by

incommensurability and conceptual discontinuities (i.e., radical semantic conversions, making impossible translating the terms of revised theories in vocabulary of previous theories);

(2) Time order: succession of cumulative and revolutionary phases, replacing each other at certain intervals of time.

Then refutation of this theory can go in one of the following ways:

(i) To demonstrate conceptual continuities in revolutionary science
(ii) To demonstrate the noncumulative character of normal science
(iii) To demonstrate the absence of sequential change of phases in time through the coexistence of competing theories as a usual fact of scientific practice

And (i) is more effective than (ii) because the conceptual change within normal science, if radical, can be easily interpreted as revolutionary or, otherwise, as minor. Demonstrating lack of support for incommensurability, Mizrahi (2015) exposes considerations, which correspond to the strategy in (i). Again, Kuhn's argument can be schematized as follows:

(1) If there are revolutions, then there must be normal science.
(2) x, y, and z are examples of scientific revolutions.
(3) Normal science exists.
(4) The theory of scientific revolutions is correct.

Let us suppose that Kuhn did provide convincing historical examples of scientific revolutions. The revolutions sparked by Copernicus, Newton, Einstein, and Dalton seem to satisfy Kuhn's descriptions for, at least, the incommensurability of key concepts. The meanings of notions such as "planet," "earth," "sun," "moon," "motion," "mass," "space," "time," and "matter" were converted radically in the theories of these scientists. If premises (1) and (2) are correct, conclusions (3) and (4) are also correct. The counter-argument to Kuhn's argument may be schematized as follows:

(1) If there are conceptual continuities during revolutionary theory-change, then

(a) revolutionary science does not exist, or;
(b) revolutionary science does not necessarily presuppose conceptual discontinuities, but;
(c) revolutionary science necessarily presupposes conceptual discontinuities;

(2) *x*, *y*, and *z* are examples of conceptual continuities in revolutionary science;

(3) revolutionary science does not exist (and then see Kuhn's argument discussed earlier);

(4) The theory of scientific revolutions is not correct.

Demonstrating lack of deductive support for the incommensurability thesis, Mizrahi justifies our inclusion of (b) in premise (1). Mizrahi argues that any conceptual change does not necessarily presuppose lexical incommensurability. Shifts from one theory to another can be accompanied by commensurable and compatible conceptual revisions: "there are no conclusive epistemic reasons to think that scientific change, revolutionary or otherwise, involves abandoning a lexical taxonomy in favour of another incompatible one, given that reference change alone is not conclusive evidence for incompatibility of conceptual content" (Mizrahi 2015, 365). But Mizrahi's objections do not work against Kuhn, even if reasonable in general. According to Kuhn, incommensurability forms the foundation for the idea of revolutions, and both concepts are interconnected by definition:

> One aspect of every revolution is, then, that some of the similarity relations change. Objects which were grouped in the same set before are grouped in different sets afterwards and vice versa (Kuhn 1970b, 275).
>
> When writing the book on revolutions, I described them as episodes in which the meanings of certain scientific terms changed, and I suggested that the result was an incommensurability of view-points and a partial breakdown of communication between the proponents of different theories (Kuhn 1977, xxii).

That is why (c) is included in premise (1). Incommensurability is not something Kuhn proves but the starting point in his argument. It constitutes the integral part of the theoretical lens through which scientific practice gets interpreted.

When Mizrahi demonstrates lack of inductive support for the incommensurability thesis, he adds to the justification of premise (2) in the counterargument discussed earlier in the chapter by giving the examples of conceptual change without discontinuities. Let us suppose that Mizrahi and others did provide convincing examples of conceptual continuities during revolutionary shifts in fundamental theories. Relevant cases come from the history of biology. Mayr (1972) and Greene (1971) explained in detail how Darwinian ideas about biological development were connected with previous evolutionary views. The concept of "gene," when first proposed by Johannsen, meant an element of heredity matter absent a further empirical elaboration. Earlier, this abstract idea was incorporated in various views on heredity, which shared the core idea about the existence of particles responsible for transmitting traits from one organism to another. In modern genetics, this basic understanding of "gene" did not change, but the initial concept was essentially concretized in relation to the gene-trait relation, functions, material basis, and structure of genes.[5] And I suppose that additional similar evidence can be found in the history of science. If the premises in the counter-argument are correct, the conclusions are also correct.

Something must be wrong with the premises and the logic of Kuhn's theory if contradictory conclusions are derived from the two arguments discussed earlier in the chapter. The notion of scientific revolution implies change on a fundamental level, characterized by incommensurability and discontinuities. Mizrahi rightly points out that Kuhn does not give conclusive epistemic reasons why incommensurability must accompany the change of a theoretical framework. That is just a choice Kuhn makes to understand revolutionary change in this way, which may not be upheld. It is always possible to prefer other ways. This breaks Kuhn's argument but opens the possibility for alternative understanding of revolutions and conceptual change, which takes into account divergent cases with conceptual continuities and discontinuities, found in historical studies of scientific development.

The theoretical framework of theories evolves and changes constantly. Critics were right in saying that the distinction between normal and revolutionary science hardly "holds water" (Toulmin 1970). Still, radical changes in fundamental theories and research practices can be marked as revolutionary if "revolution" is understood as the label that scientists, historians, or philosophers apply to the significant periods in the history of science, regardless as to whether they display any kind of continuity or discontinuity in relation to previous knowledge. The label "revolution" can be understood like an award or a sign of collective appraisal for distinct developments in science.

As for the absence of sequential change of normal and revolutionary phases, Mayr (1972) and Greene (1971) demonstrated as much regarding the historical material of biology, in particular for the conception of evolution, which competed with various evolutionist and anti-evolutionist views for almost a century before and after the crucial work on this idea written by Darwin. Interesting examples are found by Jacobson in the history of neuroscience, in particular for neuron theory: "The greatest 'revolution' in neuroscience is conventionally taken to be the overthrow of the reticular theory of nerve cell connections by the neuron theory, but . . . elements of both theories coexisted for at least 70 years, which is almost half the duration of the history of modern neuroscience" (Jacobson 1993, 35). In the social sciences, the coexistence of competing theories and approaches is the usual state of affairs (Burrell 2005), though it is questionable whether Kuhn's lens is designed for fields in the social and human sciences (this theme will be discussed further).

Revolutions, in the Kuhnian sense, bring together radical disruptions in conceptual frameworks, metaphysics, methodologies, and values. Revolutions, then, are rare events. Fundamental changes do not occur suddenly as a switch between different paradigms, but proceed through periods of preparation, discussion, and development.[6] The distinction between puzzle-solving normal science and innovative revolutionary science disappears if the constantly developing character of theoretical frameworks and coexistence of alternatives are recognized. Still, distinct breakthroughs in science can be marked as revolutions, but no universal system of criteria for such appraisal can be formulated in a normative philosophical manner.

Limitations

The theory of scientific revolutions serves as a universal explanatory model of scientific development, potentially apt to the social and human sciences, as well as the natural sciences. On the one hand, Kuhn (1970a, 15) said that "it remains an open question what parts of social science have yet acquired such paradigms at all. History suggests that the road to a firm research consensus is extraordinarily arduous." He was unsure whether these fields have the established paradigms that develop according to the logic of his theory. On the other hand, Kuhn tried to fit social and human sciences to his model, explaining the lack of consensus on fundamentals as their immaturity: "In any case, there are many fields—I shall call them protosciences—in which practice does generate testable conclusions but which nonetheless resemble philosophy and the arts rather than the established sciences in their developmental patterns. I think, for example, [. . .] of many of the social sciences today" (Kuhn 1970b, 244).

Protosciences belong to the preparadigm phase of scientific growth, when there is no single pattern for conducting research and various approaches compete with one another. For Kuhn, socio-humanistic studies form immature science at the preparadigm phase rather than areas with specific logic of development. As Kuhn (2000, 222–23) writes:

> I'm aware of no principle that bars the possibility that one or another part of some human science might find a paradigm capable of supporting normal, puzzle-solving research. And the likelihood of that transition's occurring is for me increased by a strong sense of déjà vu. Much of what is ordinarily said to argue the impossibility of puzzle-solving research in the human sciences was said two centuries ago to bar the possibility of a science of chemistry and was repeated a century later to show the impossibility of a science of living things. Very probably the transition I'm suggesting is already under way in some current specialties within the human sciences. My impression is that in parts of economics and psychology, the case might already be made.

In the introduction to *Structure*, the methodology of research is bound with history and sociology of science where philosophical generalizations and interpretations are rooted and tested with empirical case studies (Kuhn 1970a, 8–9). Kuhn (1970a, 4) explicitly characterizes his own method as follows: "an analysis of the development of scientific knowledge must take account of the way science has actually been practiced." Kuhn considered himself as a historian no less than as a philosopher of science. However, the philosophical nature of his theory is more manifest. Its universalist ambitions for explaining science resemble the search for a universal methodology of cognition in classical epistemology. That is why Feyerabend argued against Kuhn as follows:

> his general approach confused many writers: finding that science had been freed from the fetters of a dogmatic logic and epistemology they tried to tie it down again, this time with sociological ropes. That trend lasted well into the early seventies. By contrast there are now historians and sociologists who concentrate on particulars and allow generalities only to the extent that they are supported by sociohistorical connections (Feyerabend 1993, x).

The very existence of the phenomenon of science suggests the idea that science could be described with general features despite the differences among distinct fields, subjects, and approaches within it. The range of modern conceptions (e.g., technoscience, big and small science, and postacademic science) is created on the same level of generalization. So, what must be mistaken about the universalist ambitions of Kuhn's model if general statements about scientific practice and history are meaningful and acceptable in numerous contexts?

The conceptions of technoscience, big and small science, or postacademic science are based on generalizations about past and present scientific practices. They include descriptions and periodizations but not predictions about future of scientific development. Meanwhile, Kuhn's theory is supposed to be applicable to the past, present, and future of science. It searches for the universal logic of scientific growth and laws of history. It bears the dubious philosophical presupposition that the universal nature of scientific development can be extracted from the retrospective analysis of the history of science. Consequently, its limitations became more obvious over time. How greatly dependent on the historical moment Kuhn's statement seems today: "Science is not the only activity the practitioners of which can be grouped into communities, but it is the only one in which each community is its own exclusive audience and judge" (Kuhn 1970b, 254). Inter- and trans-disciplinary communications in modern science expanded the audience of scholars. Moreover, scientific communities invited Internet users into their ranks. The emergence of crowd-sourced science in the twenty-first century opens the possibility for the collaboration of scientists with the public in all stages of research including organizing, funding, data searching, analyzing, generating hypotheses, testing, and evaluating results. It is impossible to predict exactly how scientific research will be changed tomorrow by the progress of information technology and artificial intelligence. There are no decisive reasons even to think that science will remain among the primary occupations of humans in future.

In large part, this lesson was learned from the critique of law-like generalizations in historical and philosophical thinking. Different aspects of the theory of scientific revolutions are marked as controversial but not overcome in contemporary science studies. Modeling scientific development, Kuhn appeals to the history and practice of physical science. According to his assumption, the logic of the development of other sciences must be the same. Yet, the specific characteristics of other sciences, especially the human sciences, are largely ignored by such an approach. It feeds the old philosophical dogma that fields that do not accord with the standards and methods of physics must be considered as protosciences, that is, immature and defective in some sense. Even if the relevance of an account in the human sciences appears particularly significant, studies in the philosophy of science continue to focus mostly on the practices of the natural sciences, taking them as the basis for general conclusions about all science.

The actual discourses in science and technology studies (STS), sociology, and philosophy of science focus significantly on the themes of technology, technoscience, and technoculture. And these themes are analyzed usually from the standpoint of the state of affairs in the natural sciences. At the same time, conceptions of technoscience and technoculture contain axiological implications that influence the development of science and culture. Giving priority to the natural sciences, they set the policy, which downplays the significance of human sciences for society. In STS, sociology, and philosophy of science, there is a tendency for reification of technologies (Barnes 2005, 146), that is, understanding them as material objects, artifacts, mechanisms, and technical devices, developed on the basis of knowledge in natural sciences. Another way of understanding can be produced from antique usage of the word *techne*, which envisages the broader meaning for technology as mastery, art, professional skills, and methods of implementation of certain kinds of activity.

Following the old tradition of usage, it is possible to reconceptualize the ideas of technology, technoscience, and technoculture to the extent that they unequivocally embrace social and human technologies based exclusively on socio-humanistic knowledge, practices of social engineering and planning, and various practical dimensions of modern human sciences (Argamakova 2017). In that case, the language of technoscientific discourse would become more suitable for describing the cognitive practices of social and human sciences, which cannot be done for many other theories of science, displaying their limitations when explaining the actual state of affairs and history of these fields. This bias in favor of natural science is shown by the cases of technoscience as well as the theory of scientific revolutions. There are more social factors that are responsible for its existence than cognitive reasons for accepting it. Those benefits, such as the natural knowledge brought to society, determine its special place in the system of cognitive practices. However, a more comprehensive analysis of science requires equal attention to the specific aspects of the human sciences and recognition of their significance.

We may well ask why we continue to search for models of science if previous generalizations were limited, ambiguous or inaccurate. Perhaps it is a peculiar ambition to try to describe science's general features. Finding so many flaws in Kuhn's theory, one wonders how his ideas attracted enormous attention. Kuhn's work, at least, provokes these questions about science. The history of philosophy shows that the best intellectual provocations are valuable in themselves. Their merit consists not in the answers on all possible questions, but in the variety of the lessons we learn in order to move knowledge forward.

NOTES

1 See Academy of Arts and Sciences (1996), Lackey (1999), and Abbot (2016).

2 This is the wave of critical writings that appeared in ten years following the publication of *Structure*, including Hesse (1963), Shapere (1964), Stopes-Roe (1964), Scheffler (1967), Lakatos and Musgrave (1970), Ruse (1970) and (1971), Greene (1971), and Mayr (1972).

3 Research both critical and developing Kuhn's views includes: Nickles (2002); Soler, Sankey, and Hoyningen-Huene (2008); Andersen, Barker, and Chen (2006); Vosniadou (2008).

4 See Barnes (1977) and (1982).

5 Weber (2005) applied the conception of reference potential to explain the sort of semantic continuity through episodes of conceptual revisions which the concept of "gene" experienced. In general, he argued that the meaning of "gene" was neither stable nor unstable, but floating.

6 The close ideas can be found in, for example, Vosniadou (2008).

REFERENCES

Abbot, Andrew. 2016. "*Structure* as Cited, *Structure* as Read." In *Kuhn's Structure of Scientific Revolutions at Fifty: Reflections on a Science Classic*, edited by Robert J. Richards and Lorraine Daston, 167–81. Chicago: University of Chicago Press.

Academy of Arts and Sciences. 1996. "The Hundred Most Influential Books since the War." *Bulletin of the American Academy of Arts and Sciences* 49, 12–18.

Andersen, Hanne, Peter Barker, and Xiang Chen. 2006. *The Cognitive Structure of Scientific Revolutions*. Cambridge: Cambridge University Press.

Argamakova, Alexandra. 2017. "Social and Humanitarian Dimensions of Technoscience." *Epistemology and Philosophy of Science* 52, 120–36.

Barnes, Barry. 2005. "Elusive Memories of Technoscience." *Perspectives on Science* 13, 142–65.

Barnes, Barry. 1977. *Interests and the Growth of Knowledge*. London and New York: Routledge.

Barnes, Barry. 1982. *T. S. Kuhn and Social Science*. London: Macmillan.

Burrell, Gibson. 2005. *Sociological Paradigms and Organisational Analysis: Elements of the Sociology of Corporate Life*. Ardershot: Ashgate Publishing.

Feyerabend, Paul. 1993. *Against Method*. London: Verso.

Greene, John C. 1971. "The Kuhnian Paradigm and the Darwinian Revolution in Natural History." In *Perspectives in the History of Science and Technology*, edited by D. H. D. Roller, 3–25. Norman: University of Oklahoma Press.

Hacking, Ian. 2012. *Introductory Essay to* The Structure of Scientific Revolutions *by Thomas Kuhn*, vii–xxxvii. Chicago: The University of Chicago Press.

Hesse, Mary. 1963. "Review of *The Structure of Scientific Revolutions*." *Isis* 54, 286–87.

Jacobson, Marcus. 1993. *Foundations of Neuroscience*. New York: Plenum Press.

Kuhn, Thomas. 1970a. "Logic of Discovery or Psychology of Research?" In *Criticism and the Growth of Knowledge*, edited by Imre Lakatos and Alan Musgrave, 1–23. Cambridge: Cambridge University Press.

Kuhn, Thomas. 1970b. "Reflections on My Critics." In *Criticism and the Growth of Knowledge*, edited by Imre Lakatos and Alan Musgrave, 231–78. Cambridge: Cambridge University Press.

Kuhn, Thomas. 1970c. *The Structure of Scientific Revolutions*. Chicago: University of Chicago Press.

Kuhn, Thomas. 1977. *The Essential Tension. Selected Studies in Scientific Tradition and Change*. Chicago: University of Chicago Press.

Kuhn, Thomas. 2000. *The Road Since Structure*. Chicago: University of Chicago Press.

Lackey, Douglas P. 1999. "What Are the Modern Classics? The Baruch Poll of Great Philosophy in the Twentieth Century." *The Philosophical Forum* 30, 329–46.

Lakatos, Imre, and Alan Musgrave, eds. 1970. *Criticism and the Growth of Knowledge*. Cambridge: Cambridge University Press.

Masterman, Margaret. 1970. "The Nature of a Paradigm." In *Criticism and the Growth of Knowledge*, edited by Imre Lakatos and Alan Musgrave, 59–89. Cambridge: Cambridge University Press.

Mayr, Ernst. 1972. "The Nature of Darwinian Revolution." *Science* 176, 981–89.

Mizrahi, Moti. 2015. "Kuhn's Incommensurability Thesis: What's the Argument?" *Social Epistemology* 29, 361–78.

Nickles, Thomas, ed. 2002. *Thomas Kuhn*. Cambridge: Cambridge University Press.

Richards, Robert J., and Lorraine Daston. 2016. "Introduction." In *Kuhn's Structure of Scientific Revolutions at Fifty: Reflections on a Science Classic*, edited by Robert J. Richards and Lorraine Daston, 1–11. Chicago: University of Chicago Press.

Ruse, Michael. 1970. "The Revolution in Biology." *Theoria* 36, 1–22.

Ruse, Michael. 1971. "Two Biological Revolutions." *Dialectica* 25, 17–38.

Sankey, Howard. 1993. "Kuhn's Changing Concept of Incommensurability." *The British Journal for the Philosophy of Science* 44, 759–74.

Scheffler, Israel. 1967. *Science and Subjectivity*. Indianapolis: Bobbs-Merrill.

Shapere, Dudley. 1964. "Review of *The Structure of Scientific Revolutions*." *Philosophical Review* 70, 383–94.

Soler, Lena, Howard Sankey, and Paul Hoyningen-Huene, eds. 2008. *Rethinking Scientific Change and Theory Comparison* (Boston Studies in the Philosophy of Science 255). Dordrecht: Springer.

Stopes-Roe, H. V. 1964. "Review of *The Structure of Scientific Revolutions*." *British Journal for the Philosophy of Science* 15, 158–61.

Toulmin, Stephen. 1970. "Does the Distinction Between Normal and Revolutionary Science Hold Water?" In *Criticism and the Growth of Knowledge*, edited by Imre Lakatos and Alan Musgrave, 39–47. Cambridge: Cambridge University Press.

Vosniadou, Stella, ed. 2008. *International Handbook of Research on Conceptual Change*. New York: Routledge.

Weber, Marcel. 2005. *Philosophy of Experimental Biology*. Cambridge: Cambridge University Press.

Chapter 3

Can Kuhn's Taxonomic Incommensurability Be an Image of Science?

Seungbae Park

I criticize Kuhn's (1962/1970) taxonomic incommensurability thesis as follows: (i) His argument for it is neither deductively sound nor inductively correct. (ii) It clashes with his account of scientific development that employs evolutionary theory. (iii) Even if two successive paradigms are taxonomically incommensurable, they have some overlapping theoretical claims, as selectivists point out. (iv) Since scientific revolutions were rare in the recent past, as historical optimists observe, they will also be rare in the future. Where scientific revolution is rare, taxonomic incommensurability is rare, and taxonomic commensurability is common. For these reasons, taxonomic commensurability rather than incommensurability should be advanced as an image of science.[1]

1. OVERVIEW

Thomas Kuhn (1962/1970, 149–50) famously claims that competing paradigms are taxonomically incommensurable. I call Kuhn's claim "the taxonomic incommensurability thesis (TI)," following Moti Mizrahi (2015, 362). It appears that taxonomic incommensurability can be advanced as an image of science, when combined with Kuhn's view that scientific development consists of cycles of normal science and revolutionary science. (TI) and the account of scientific development jointly imply that scientific revolutions will occur in the future as they did in the past, and as a result, present paradigms will be displaced by taxonomically incommensurable new ones. Thus, taxonomic incommensurability is a perennial phenomenon in science.

This chapter aims to show that taxonomic incommensurability cannot be advanced as an image of science. I proceed as follows. In section 2, I delineate what a deductively sound argument and an inductively correct argument for (TI) would look like. I also argue that it is impossible to construct a deductively sound argument for (TI), and that it is difficult, although possible, to construct an inductively correct argument for (TI). In section 3, I argue that Kuhn's argument for (TI) is neither deductively sound nor inductively correct. In section 4, I argue that (TI) clashes with Kuhn's contention that science evolves in the way that organisms do. In section 5, I show that the taxonomic incommensurability of successive paradigms does not mean that an old paradigm is thrown out *in toto*. In section 6, I argue that scientific revolutions will be rare in the future, as they have been in the recent past, and hence that taxonomic incommensurability will rarely arise in the future. In section 7, I anticipate and reply to some possible objections.

2. DEDUCTIVELY SOUND ARGUMENT AND INDUCTIVELY CORRECT ARGUMENT

Mizrahi (2015, 363–68) raises and answers the question: Can there be a deductively sound argument for (TI)? I wish to raise and answer a related question: If there were a deductively sound argument for (TI), what would it look like?

Let me use an analogy to answer this question. Suppose there are some balls in an urn. You believe that all of them are red. How would you go about constructing a deductively sound argument for your belief? The answer to this question is simple and straightforward. You pull out all the balls from the urn and then check each ball for its color. If there are ten balls and all of them are red, you can construct a deductively sound argument as follow:

> There are ten balls in the urn.
> Ball$_1$, ball$_2$ . . . and ball$_{10}$ are red.
> ∴ All the balls in the urn are red.

In this argument, the conclusion necessarily follows from the premises, and the premises are true. So the argument is deductively sound.

This example illustrates how we can go about constructing a deductively sound argument for (TI). Suppose that there are 100 pairs of competing paradigms in science, that you check each pair for taxonomic incommensurability, and that all of them are taxonomically incommensurable. You can then construct a deductively sound argument for (TI) as follows:

> There are one hundred pairs of competing paradigms in science.
> $Pair_1$, $pair_2$ . . . and $pair_{100}$ are taxonomically incommensurable.
> ∴ All pairs of competing paradigms are taxonomically incommensurable.

Can we construct such an argument for (TI)? My answer is no. Constructing such an argument would require us to enumerate all the pairs of rival paradigms in science, and to check each pair for taxonomic incommensurability. The set of all the pairs includes not only past, but also future, paradigms. But how can we be certain about the number and the contents of future rival paradigms? Set this problem aside. There is another. As Marc Lange (2002, 283) points out, it is difficult to individuate and count scientific theories. It follows that it is also difficult to individuate and count paradigms. Kuhn (1962/1970) does not enumerate past paradigms. He does not even say how many paradigms there were in the history of science. He only says that "the examples could be multiplied *ad nauseam*" (1962/1970, 136). It is not surprising why he only says so, given that it is difficult to individuate and recognize paradigms.

Mizrahi (2015, 368–73) raises and answers the question: Can there be an inductively correct argument for (TI)? I wish to raise and answer a related question: If there is an inductively correct argument for (TI), how can it be constructed?

My answer to this question is that some pairs of competing paradigms should be randomly selected from the set of all the pairs of competing paradigms. If not random, the fallacy of biased statistics would occur. In addition, the number of selected pairs should be large enough. If not, the fallacy of hasty generalization would arise. Only if these two conditions were met, would the sample be representative of the general population of the pairs of successive paradigms and the inference from the sample to the population be inductively correct. It would be possible, although difficult, to construct such an argument, once we set aside the problem of selecting future paradigms and the problem of individuating and recognizing paradigms.

In this section, I delineated what a deductively sound argument and an inductively correct argument for (TI) would look like, and whether and how they can be constructed. In the next section, I show that Kuhn's (1962/1970; 2000) argument for (TI) is neither deductively sound nor inductively correct.

3. KUHN'S ARGUMENT

Kuhn (1962/1970) offers two examples to argue for (TI). One example concerns Newtonian and Einsteinian mechanics. Under Newtonian and

Einsteinian mechanics, space is unaffected and affected by the presence of matter, respectively, so the meaning of "space" changed (Kuhn, 1962/1970, 149). His other example concerns Ptolemaic and Copernican astronomy. Under Ptolemaic and Copernican astronomy, the Earth does not move and moves, respectively, so the meaning of "the Earth" changed (1962/1970, 149).

Kuhn (2000) offers three examples to argue for (TI). The first example involves the concept of motion in Aristotelian and Newtonian mechanics. In Aristotelian mechanics, the term "motion" refers to change in general, whereas in Newtonian mechanics, it refers only to a change of positions (2000, 17). The second example involves the historical episode that the concept of a cell changed as a result of the replacement of the contact theory of a battery with the chemical theory of battery (2000, 20–24). The third example involves Max Planck's replacement of the terms "energy element" and "resonator" with the new terms "energy quantum" and "oscillator" (2000, 24–28).

In total, Kuhn (1962/1970; 2000) uses five examples to argue for (TI). He does not claim that they are exhaustive, so his argument for (TI) is not deductively sound. In addition, he claims neither that the five examples are randomly chosen from the population of successive paradigms nor that the number of the examples is large enough. So his argument for (TI) is not inductively correct. In short, he has offered neither a deductively sound argument nor an inductively correct argument for the position that all pairs of competing paradigms are taxonomically incommensurable.

Mizrahi is on the right track when he says that "it is a mistake to generalize from a few selected examples that competing theories in general are taxonomically incommensurable" (2015, 368). Kuhn's argument for (TI) shows, at best, that five pairs of consecutive paradigms are taxonomically incommensurable. It must be independently argued that other pairs of successive paradigms in science, such as the pair of the caloric and the kinetic paradigm and the pair of the ether and the electromagnetic paradigms, are also taxonomically incommensurable.

It might be, however, that Kuhn does not go into the details of other pairs of competing paradigms in science to save space. So I will grant, for the sake of argument, that all paradigms before the early twentieth century were taxonomically incommensurable with their successors. Even so, I argue in the following sections, it is problematic to advance taxonomic incommensurability as an image of science.

4. EVOLUTIONARY THEORY

Kuhn contends that science does not move toward a goal, just as organisms do not evolve toward a goal, and that the analogy between the evolution of science and that of organisms is "very nearly perfect" (1962/1970, 172). Thus, it is wrong to think that science moves toward truths. First, note that this claim about the evolution of science and that of organisms presupposes that evolutionary theory is true. After all, if evolutionary theory is false, then so would be his claim about the evolution of science (Park 2017a, 324–325).

Recall, however, that according to Kuhn, scientific development consists of cycles of normal and revolutionary science. This account of scientific development and (TI) jointly imply that evolutionary theory will be superseded by a taxonomically incommensurable new theory, and hence that Kuhn's account of scientific development will also be superseded by a taxonomically incommensurable new account of scientific development. So, for example, what is meant by "paradigm" under Kuhn's account of scientific development will be different from what is meant by the term "paradigm" under the new account. Just as the word "planet" picks out different objects under the Ptolemaic and Copernican paradigms, "paradigm" will pick out different segments of science under Kuhn's account of scientific development and its successor. Or perhaps according to the new concept of paradigm, competing paradigms will be commensurable, and a scientific revolution will be completed when a group of scientists persuades another group of scientists through rational argumentations. Or it might be that future philosophers will use different terms, as Imre Lakatos (1970) does, to give an account of scientific development.

Kuhn might reply that evolutionary theory is an exception to his account of scientific development and (TI), so evolutionary theory will not be replaced with a taxonomically incommensurable new theory. So his account of scientific development and (TI) will also not be replaced by a new account of scientific development. In other words, evolutionary theory and his account of scientific development will remain stable unlike our other best theories, such as the kinetic theory and the special theory of relativity.

It is not clear, however, how plausible this move is. What is so special about evolutionary theory that sets it apart from our other best theories? What is the reason for thinking that evolutionary theory will not be superseded by a taxonomically incommensurable alternative, while our other best theories will be? Without convincing answers to these questions, it is merely ad hoc to say that evolutionary theory is an exception to Kuhn's account of scientific development and (TI).

Kuhn might reply that he does not need evolutionary theory *in toto* to defend his account of science. His account of scientific development only requires the observational claim of evolutionary theory that organisms do not evolve toward a goal. Scientific claims can be divided into theoretical and observational claims. Antirealists can avail themselves of observational claims, but not theoretical claims, to develop an account of scientific development. So it is not necessarily incoherent to appeal to evolutionary theory to give an account of scientific development, which discredits the theoretical claims of science.

This possible reply from Kuhn, however, faces the following two objections. First, the claim that organisms do not evolve toward a goal is predicated on the claim that organisms have not evolved toward a goal since the beginning of life about four billion years ago. It is controversial whether such a claim is observational or theoretical. Although it can be classified as observational by fiat, it cannot be directly confirmed in the way an observational claim that a cat is on the mat is directly confirmed. We can only infer to it from other observational claims. It is also not clear whether it can be better confirmed than theoretical claims, such as the claim that the motion of molecules is responsible for heat.

Second, appealing to Gestalt psychology, Kuhn (1962/1970, 111–35) advances the famous claims about observation, for example, observation is theory-laden, scientists of competing paradigms live in different worlds, and observational data cannot serve as neutral arbiters between competing paradigms. I set aside the issue of whether Gestalt psychology is supported by theory-laden or theory-neutral data. I instead pursue the issue of whether or not we can trust the observational claim that organisms do not evolve toward a goal. If Kuhn is right that observation is theory-laden, then the observational claim that organisms do not evolve toward a goal is also contaminated by evolutionary theory. Hence, it will not be endorsed by future scientists working under a different paradigm. Philosophers of science who will invoke the alternative paradigm will also disagree with Kuhn about whether science moves toward a goal or not. These philosophers and Kuhn, if alive, will live in different worlds, and observations about science will not be able to serve as neutral arbiters between them. A moral here is that it is self-defeating for Kuhn to invoke a scientific theory to give an account of science that discredits scientific claims, theoretical and observational.

5. SELECTIVISM

Concerning Newton's second law of motion, Kuhn claims that the "concepts of force and mass deployed in that law differed from those in use before the law was introduced" (2000, 15). Based on this observation, he claims that when "referential changes of this sort accompany change of law or theory, scientific development cannot be quite cumulative" (2000, 15). In other words, he claims that scientific development cannot be cumulative due to taxonomic incommensurability. So proponents of (TI) might be tempted to think that a taxonomically new paradigm ousts an old paradigm *in toto*.

A closer look into the history of science, however, reveals that it is wrong to think so. As John Worrall (1989), Philip Kitcher (1993, 140–49), Jarrett Leplin (1997), and Stathis Psillos (1999, chapters 5 and 6) point out, when theories were ousted, not all theoretical claims of an old theory were thrown out; some theoretical claims were carried over to the new theory. Their position is called selective realism or selectivism. The idea is that we should be selective about which theoretical claims are worthy of our belief and which are not.

Selectivism applies to the very examples that Kuhn uses to argue for (TI). Consider the transition from the Ptolemaic theory to the Copernican theory. Some theoretical claims of the Ptolemaic theory were retained in the Copernican theory, such as the claim that the orbit of Mars falls inside that of Jupiter, and the claim that the orbit of Jupiter falls inside that of Saturn. Consider also the transition from Newtonian mechanics to Einsteinian mechanics. As Michael Friedman (2001, 63) observes, some theoretical claims of Newtonian mechanics are enshrined in Einsteinian mechanics. For example, Euclidean geometry and the law of inertia were carried over from classical mechanics to the special theory of relativity.

Selectivists would agree with Kuhn that the Sun is classified as a planet under the Ptolemaic theory, whereas it is classified as a star under the Copernican theory, and that $m=F/a$ in Newtonian mechanics, whereas, $m=E/c^2$ in Einsteinian mechanics. They would point out, however, that the Ptolemaic theory and Newtonian mechanics shared some theoretical assumptions with the Copernican theory and Einsteinian mechanics, respectively. Thus, the rival paradigms had some overlapping assumptions about unobservables, although the paradigms were taxonomically incommensurable.

Psillos draws two interesting conclusions from the fact that past and present theories have some overlapping assumptions about unobservables. First, he concludes that past theories were not completely false but approximately true (1999, 113). Second, he concludes that past theoretical terms like "phlogiston" and "ether" *approximately* refer to the referents of present theoretical terms like "oxygen" and "electromagnetic field" (1999, 294). On this account, reference admits of degrees; it is not an all-or-nothing affair. His theory of reference strengthens his contention that past theories were approximately true and weakens (TI) which presupposes that reference is an all-or-nothing affair.

How do scientific antirealists criticize selectivism? P. Kyle Stanford (2015, 876) claims that there is only a terminological dispute between realists and antirealists. Realists affirm, while antirealists deny, that the preserved theoretical claims of past theories are rich enough to attribute "approximate truths" to past theories. Stanford's observation indicates that not all theoretical constituents of past theories were abandoned. Thus, realists and antirealists alike would deny that, if rival paradigms are taxonomically incommensurable, then they have no overlapping theoretical assumptions, or that a new paradigm will oust all the theoretical assumptions of an old paradigm.

6. HISTORICAL OPTIMISM

Scientific realists have developed various theoretical resources to defuse the pessimistic induction that since past theories were abandoned, present theories will also be abandoned. Selectivism is one of them. Another was developed by Ludwig Fahrbach (2011a, 148), Park (2011, 79), and Mizrahi (2013, 3220). They distinguish between distant and recent past theories. Distant past theories include the Ptolemaic theory, the humoral theory, the phlogiston theory, the caloric theory, and the ether theory. These theories were all abandoned before the early twentieth century. Recent past theories include the germ theory, the oxygen theory, the kinetic theory, and the special theory of relativity. They were accepted in the twentieth century and have not yet been rejected. Since they are still accepted in the early twenty-first century, they can also be regarded as present theories. The set of recent past theories is far larger than that of the distant past theories. Fahrbach, for instance, observes that "at least 95% of all scientific work ever done has been done since 1915, and at least 80% of all scientific work ever done has been done since 1960" (2011a, 148). Fahrbach, Park, and Mizrahi's observation of the history of science is dubbed "historical optimism" (Park 2017b, 316). Historical optimism rebuts the premise of the pessimistic induction that all (or most) past theories were rejected.

Historical optimism also rebuts (TI). Even if all distant past paradigms were taxonomically incommensurable with their successors, for example, with recent past paradigms, it is still problematic to say that most past paradigms were taxonomically incommensurable with their successors. Most past paradigms were recent past paradigms, and recent past paradigms have no successors yet. If we randomly select some paradigms from the general population of past paradigms, most of the selected paradigms would be recent past paradigms. Mizrahi (2013, 3219–20) has already carried out such a random sampling, selecting 40 theories out of 124 past theories. It turns out that 29 of them were still accepted theories, 6 were abandoned theories, and 5 were debated theories. Thus, to argue that most past paradigms were taxonomically incommensurable with their successors on the basis of some examples of distant past paradigms is to commit the fallacy of biased statistics. In sum, historical optimism forestalls any attempt to construct an inductively correct argument for (TI).

Historical optimism goes hand in hand with Kuhn's examples of paradigms. It is a tricky business to individuate and recognize paradigms, as we noted earlier, but Kuhn uses the following examples of paradigms throughout his book (1962/1970):

Kuhn's Examples of Paradigms

- Ptolemaic astronomy, Copernican astronomy (Kuhn 1962/1970, 10)
- Phlogiston chemistry (2), the oxygen theory of combustion (56)
- Aristotelian dynamics (10), the scholastic impetus paradigm (120),
- Newtonian dynamics (10), Einsteinian dynamics (110), quantum mechanics (49)
- Franklinian paradigm of electricity (18)
- Catastrophism, Uniformitarianism (48)
- Newton's corpuscular optics, Young and Fresnel's wave optics (12), Maxwell's electromagnetic theory (58), Einstein's corpuscular optics, the quantum optics (12)
- Caloric thermodynamics (2, 29), statistical mechanics (48)
- Pre-Darwinian evolutionary theories (171), Darwin's theory of evolution (20, 151, 171)
- Affinity theory, Dalton's atomic theory (131)

Kuhn does not use other examples of paradigms in his book (1962/1970). Compare Kuhn's list of paradigms with Stanford's (2006, 19–20) list of transitions from past to present theories:

Stanford's List

- from elemental to early corpuscularian chemistry to Stahl's phlogiston theory to Lavoisier's oxygen chemistry to Daltonian atomic and contemporary chemistry
- from various versions of preformationism to epigenetic theories of embryology
- from the caloric theory of heat to later and ultimately contemporary thermodynamic theories
- from effluvial theories of electricity and magnetism to theories of the electromagnetic ether and contemporary electromagnetism
- from humoral imbalance to miasmatic to contagion and ultimately germ theories of disease
- from eighteenth-century corpuscular theories of light to nineteenth-century wave theories to the contemporary quantum mechanical conception
- from Darwin's pangenesis theory of inheritance to Weismann's germ-plasm theory to Mendelian and then contemporary molecular genetics
- from Cuvier's theory of functionally integrated and necessarily static biological species and from Lamarck's autogenesis to Darwin's evolutionary theory

An interesting common feature emerges between Kuhn and Stanford's examples. Their examples of past theories are of ones that were accepted and rejected before the early twentieth century; none of them was accepted after the early twentieth century. This common feature accords well with historical optimism that recent past theories have been relatively stable.

Utilizing historical optimism, Park (2016b, 10) constructs a pessimistic induction against pessimists. The pessimists of the early twentieth century, such as Henri Poincaré (1905/1952, 160) and Ernst Mach (1911, 17), predicted that scientific revolutions would occur and as a result, that the then present theories, for example, the aforementioned recent past theories, would be overturned. Their prediction has not accorded with the history of science. Since the pessimists of the early twentieth century were proved wrong about most of their present theories, the pessimists of the early twenty-first century will also be wrong about most of their present theories. This pessimistic induction over pessimists entails that most present paradigms will not be ousted by taxonomically incommensurable new ones, that most future paradigms will be similar to most present paradigms, and that taxonomic incommensurability will rarely arise in the future, as it has rarely arisen in the recent past.

K. Brad Wray would accept the pessimistic induction over pessimists, given that he says that "only the fate of our most recently developed theories are relevant to determining what we can expect of today's best theories" (2015, 63). In other words, if we want to know whether present theories will be surpassed by alternatives or not, we should investigate the recent past theories of the twentieth century, and not the distant past theories from before the twentieth century. Wray is right on this account. Present theories are more similar to recent than to distant past theories. So if we have to choose between distant and recent past theories in order to assess the fate of present theories, we should choose recent past theories rather than distant past theories. Since most recent past theories have been stable, most present theories will also be stable.

7. OBJECTIONS AND REPLIES

How might Kuhn respond to the aforementioned pessimistic induction over pessimists? He might argue that scientists today are doing normal science. For example, physicists today are fleshing out the general theory of relativity. They dogmatically stick to it, even if they encounter anomalies. It is therefore not surprising that there have been no scientific revolutions in the recent past. Revolutionary science, however, follows normal science by the very definition of the term "normal science." So, Kuhn might argue, scientific revolutions will occur, and present paradigms will be displaced by taxonomically incommensurable new ones.

Proponents of the pessimistic induction over pessimists, however, would object that it begs the question to apply the term "normal science" to what scientists are doing these days. Of course, they are fleshing out present paradigms. But applying the term "normal science" to current scientific activities implies that scientific revolutions will occur, since "normal science" is defined as that which is followed by revolutionary science. So we need a new term that is neutral as to whether present paradigms will be superseded by new paradigms or not. I propose that we use "ordinary science" instead of "normal science" to describe what scientists are doing these days. To say that scientists are doing ordinary science means that they are fleshing out existing paradigms, but that does not mean that scientific revolutions will either occur or not occur.

Kuhn might now argue that recent past paradigms will be ousted by future paradigms, as distant past paradigms were ousted by recent past paradigms. Hence, recent past paradigms will be taxonomically incommensurable with future paradigms, as distant past paradigms were taxonomically incommensurable with recent past paradigms.

Is this induction tenable? Many philosophers have brought up two important differences between distant and recent past theories. First, recent past theories are far more successful than distant past theories, as pointed out by Jarrett Leplin (1997, 141), Gerald Doppelt, (2007, 111; 2014), Juha Saatsi (2009, 358), Michael Devitt (2011, 292), Fahrbach (2011b, 1290), Park (2011, 80), and Mizrahi (2013). Second, scientists developed recent past theories with a view to overcoming problems that had beset distant past theories, as pointed out by Leplin (1997, 144) and Alexander Bird (2007, 108). For example, the general theory of relativity was proposed to explain the perihelion motion of Mercury, which was an anomaly to Newtonian mechanics. For these reasons, it is one thing that scientific revolutions occurred in the distant past, and it is another that they will also occur in the future. An argument that addresses these differences is required to assert that scientific revolutions will occur as they did in the distant past.

Kuhn might raise another objection to the pessimistic induction over pessimists. The fact that recent past theories have lasted for about 100 years does not indicate that they are true, for it usually takes more than a 100 years for a scientific revolution to occur. As Wray (2015, 64) observes, the Ptolemaic theory lasted for about 1,200 years, before it was superseded by the Copernican theory in the mid-sixteenth century. So in about 1,000 years, all present paradigms will be superseded by new paradigms.

As Wray (2015, 64) also observes, however, four scientific revolutions occurred in less than 120 years over the nature of light. There were four transitions from Newton's particle theory, to Fresnel's wave theory, to Maxwell's electromagnetic theory, to Einstein's particle theory, and to the quantum theory of light. If we take this as a typical scientific revolution, it takes only about thirty years for a scientific revolution to occur. On that reckoning, the best explanation of why recent past theories have been stable for about a 100 years is that they are true.

Moreover, as Fahrbach (2011a) observes, the body of scientific knowledge grows exponentially, and as Devitt (2011, 292) observes, present science uses more advanced technologies. Thus, present scientists have better means to discover anomalies to their existing paradigms than past scientists had. So enduring the tribunal of experience for a 100 years in the twentieth century has a higher epistemic value than withstanding the tribunal of experience for a 100 years, say, in the twelfth century. It does not prove very much that a theory lasted for 1,200 years before the sixteenth century, because scientific knowledge was slim, it was growing slowly, and past scientists did not have advanced technologies to test their theories.

8. CONCLUSION

I criticized (TI) as follows: Kuhn's argument for (TI) is neither deductively sound nor inductively correct. (TI) clashes with his account of scientific development, which invokes evolutionary theory. Even if two successive paradigms were taxonomically incommensurable, they have some overlapping theoretical claims, as selectivists point out. Since scientific revolutions were rare in the recent past, as historical optimists observe, they will also be rare in the future. Where scientific revolution is rare, taxonomic incommensurability is rare, and taxonomic commensurability is common. For these reasons, taxonomic commensurability rather than incommensurability should be advanced as an image of science.

NOTE

1 I thank Moti Mizrahi for the invitation to contribute to this book and for his useful comments on an earlier draft of this work.

REFERENCES

Bird, Alexander. 2009. "What Is Scientific Progress?" *Noûs* 41, 64–89.
Devitt, Michael. 2011. "Are Unconceived Alternatives a Problem for Scientific Realism?" *Journal for General Philosophy of Science* 42, 285–93.

Doppelt, Gerald. 2007. "Reconstructing Scientific Realism to Rebut the Pessimistic Meta-induction." *Philosophy of Science* 74, 96–118.

Doppelt, Gerald. 2014. "Best Theory Scientific Realism." *European Journal for Philosophy of Science* 4, 271–91.

Fahrbach, Ludwig. 2011a. "How the Growth of Science Ends Theory Change." *Synthese* 180, 139–55.

Fahrbach, Ludwig. 2011b. "Theory Change and Degrees of Success." *Philosophy of Science* 78, 1283–92.

Friedman, Michael. 2001. *Dynamics of Reason.* Chicago: University of Chicago Press.

Kitcher, Philip. 1993. *The Advancement of Science: Science Without Legend, Objectivity Without Illusions.* New York: Oxford University Press.

Kuhn, Thomas. 1962/1970. *The Structure of Scientific Revolutions.* Second Ed. Chicago: University of Chicago Press.

Kuhn, Thomas. 2000. *The Road since Structure,* edited by James Conant and John Haugeland. Chicago: University of Chicago Press.

Lakatos, Imre. 1970. "Falsification and the Methodology of Scientific Research Programmes." In *Criticism and the Growth of Knowledge,* edited by Imre Lakatos and Allen Musgrave, 91–196. New York: Cambridge University Press.

Lange, Marc. 2002. "Baseball, Pessimistic Inductions and the Turnover Fallacy." *Analysis* 62, 2881–85.

Leplin, Jarrett. 1997. *A Novel Defense of Scientific Realism.* New York: Oxford University Press.

Mach, Ernst. 1911. *History and Root of the Principle of the Conservation of Energy.* Translated by Philip E. B. Jourdain. Chicago: Open Court Publishing Company.

Mizrahi, Moti. 2013. "The Pessimistic Induction: A Bad Argument Gone Too Far." *Synthese* 190, 3209–26.

Mizrahi, Moti. 2015. "Kuhn's Incommensurability Thesis: What's the Argument?" *Social Epistemology* 29, 361–78.

Park, Seungbae. 2011. "A Confutation of the Pessimistic Induction." *Journal for General Philosophy of Science* 42, 75–84.

Park, Seungbae. 2017a. "Problems with Using Evolutionary Theory in Philosophy." *Axiomathes* 27 (3): 321–332.

Park, Seungbae. 2017b. "Why Should We Be Pessimistic about Antirealists and Pessimists?" *Foundations of Science* 22 (3): 613–625.

Poincaré, Henri. 1905/1952. *Science and Hypothesis.* New York: Dover.

Psillos, Stathis. 1999. *Scientific Realism: How Science Tracks Truth.* New York: Routledge.

Putnam, Hilary. 1975. *Mathematics, Matter and Method (Philosophical Papers, vol. 1),* Cambridge: Cambridge University Press.

Saatsi, Juha. 2009. "Grasping at Realist Straws." Review Symposium. *Metascience* 18, 355–62.

Stanford, P. Kyle. 2006. *Exceeding Our Grasp: Science, History, and the Problem of Unconceived Alternatives.* Oxford: Oxford University Press.

Stanford, P. Kyle. 2015. "Catastrophism, Uniformitarianism, and a Scientific Realism Debate that Makes a Difference." *Philosophy of Science* 82, 867–78.

Worrall, John. 1989. "Structural Realism: The Best of Both Worlds." *Dialectica* 43, 99–124.

Wray, K. Brad. 2015. "Pessimistic Inductions: Four Varieties." *International Studies in the Philosophy of Science* 29, 61–73.

Chapter 4

The Demise of the Incommensurability Thesis

Howard Sankey

The year 1962 saw the first proposal in print of the thesis of the incommensurability of scientific theories. In that year, Paul Feyerabend's paper, "Explanation, Reduction and Empiricism," and Thomas Kuhn's book, *The Structure of Scientific Revolutions*, were both published for the first time. In "Explanation, Reduction and Empiricism," Feyerabend employed the notion of incommensurability in the context of his critical analysis of the logical empiricist account of inter-theoretic reduction. In *Structure*, Kuhn ascribed a key role to incommensurability in the revolutionary transition between the theoretical frameworks that he called "paradigms." The two authors did not employ the term in precisely the same manner. But, for both authors, incommensurability was a relation that one scientific theory might bear to another.

The claim that theories or paradigms may be incommensurable met with considerable resistance. Some took incommensurability to involve a whole-sale change of meaning in the transition between theories. Such extreme semantic shift seemed to undermine the rationality of theory-choice owing to problems of inter-theoretic comparison and communication. To the extent that meaning variance entails discontinuity of reference, it was seen as a threat to a realist view of scientific progress as increase in knowledge about a shared domain of entities. Incommensurability was also taken to have consequences of a methodological nature. Standards of scientific theory appraisal were held not to be fixed or universal. Instead, standards depend upon and vary with theory or paradigm in relativistic fashion. Given such apparent implications, philosophers sought to defend the rational and progressive character of science against the challenge posed by the thesis of incommensurability.

I do not propose to revisit this earlier debate in detail here. Instead, I take my cue from the question posed by Moti Mizrahi: what is the argument for incommensurability? In fact, a number of different arguments were presented for incommensurability at various stages in the development of the idea. All were subjected to serious criticism. Despite this, the idea of incommensurability has proven to be remarkably resilient. Indeed, some authors write as if incommensurability were a phenomenon discovered by Feyerabend and Kuhn. As opposed to such authors, I hold that no cases of incommensurability have been established to exist. But, as we shall see, the matter turns on the delicate question of what is taken to count as a case of incommensurability.

In section 1, I will sketch three different ways in which the claim of incommensurability was presented. In section 2, I will indicate the main lines of critical response directed against the claim. In section 3, I will turn to the question of whether incommensurability is a phenomenon that has been shown to exist. In section 4, I will consider the issue of what is required for there to be a case of incommensurability. In section 5, I will reflect briefly on the discussion.

1. FEYERABEND AND KUHN ON INCOMMENSURABILITY

My aim in this section is to show that different arguments were given for incommensurability at different stages. I will not spell out the arguments in detail, and I will not provide an exhaustive analysis of the detailed developments of the idea. Instead, I will look at representative formulations of the idea by the two principal advocates of incommensurability. I will first consider Feyerabend's initial proposal of the idea in "Explanation, Reduction and Empiricism." Then I will consider Kuhn's original treatment of the issue in *Structure*, as well as his later taxonomic version of the idea.

In the interest of clarity, it is worth distinguishing between two forms of the incommensurability thesis.[1] According to the *semantic* form of the thesis, theories are incommensurable due to the semantic variation of the terms used by the theories. According to the *methodological* form of the thesis, theories are incommensurable due to an absence of shared standards of theory appraisal. For the most part, Feyerabend restricted his discussion to semantic incommensurability.[2] Kuhn's initial treatment of incommensurability involved both forms of incommensurability, though he later restricted it to semantic incommensurability. In this section and the following one, I will discuss both forms of incommensurability. However, in sections 3 and 4, I will restrict discussion to semantic incommensurability.

As mentioned at the outset, Feyerabend first employed the term "incommensurability" in the context of his discussion of the logical empiricist account of reduction. He sought to show that a formal account of reduction is impossible due to semantic variation between theories. He began by rejecting the empiricist idea of an independently meaningful observation language in which observational terms receive their meaning either from experience or from circumstances of use (1981b, 49–53). He adopts instead a realistic interpretation of theories on which observational terms receive their meaning from the theory that employs the terms. He then uses the example of the impetus theory to show that there are terms whose meaning depends upon the laws of a theory in such a way that they cannot be defined in the context of an opposing theory that is incompatible with those laws (1981b, 67–68).[3] From this, as well as the realistic interpretation of theories, he concludes that in the transition between such theories, the meaning of all of the terminology employed by the theories is subject to semantic variation. As a result of such semantic variation, "incommensurable theories may not possess any comparable consequences, observational or otherwise" (1981b, 93). He employs the term "incommensurable" to describe the relationship between the concepts of theories that is due to the inability to define the concepts of one theory on the basis of the other (1981b, 76).

In *Structure*, Kuhn proposed a model of scientific change based on analysis of cases drawn from the history of science. On this model, science is divided into periods of paradigm-based "normal science," which is broken at intervals by revolutionary displacement of paradigm. For Kuhn, the primary locus of incommensurability is the paradigm debate that takes place between defenders of the reigning paradigm and advocates of the new contender for the status of paradigm. For a number of reasons, the parties to the debate "fail to make complete contact with each other's viewpoints" (2012, 147). The proponents of competing paradigms do not agree on the set of scientific problems that are to be solved. Moreover, they do not possess shared standards for satisfactory problem-solution. In addition, "communication across the revolutionary divide is inevitably partial" (2012, 148). Such partial communication failure is due to semantic differences brought about by variation in conceptual apparatus across paradigms. Kuhn even suggests that scientists in competing paradigms "practice their trades in different worlds" (2012, 149). Paradigms exercise a deep influence on perception. As a result, there may be differences in the way that scientists in different paradigms perceive the world. These multiple factors "collectively" give rise to the incommensurability of competing paradigms. Given their incommensurability, the choice between competing paradigms is not "the sort of battle that can be resolved by proofs" (2012, 147).

The two accounts of incommensurability that I have just reviewed reflect the opening statements of the idea by its two principal advocates. In subsequent response to criticism, Feyerabend introduced minor modifications but did not fundamentally alter his view of the matter (see, e.g., 1981c and 1981d). By contrast, Kuhn continued to refine and develop his version of the idea throughout the remainder of his career. To illustrate, I will now sketch the taxonomic version of incommensurability, which formed part of Kuhn's mature thinking about the topic.

In work after *Structure*, Kuhn came increasingly to focus on the semantic form of incommensurability.[4] He took this to be connected with change of taxonomy. According to Kuhn, a central feature of revolutionary scientific change is change in the taxonomic systems that theories use to classify the items to which they apply.[5] Change of taxonomy may involve the redistribution of members among previously existing categories and modification of classificatory criteria, as well as the introduction of entirely new categories (cf. Kuhn 2000b, 29–30). Such taxonomic change has an impact at the semantic level. Where vocabulary is retained, taxonomic change may induce change in the meaning of the preserved terms, which may include variation of reference. Where new categories are proposed, new terms may be introduced that differ semantically from previously employed terms. Kuhn takes the taxonomic scheme of a theory to be represented by a structured vocabulary of natural kind terms, which he calls a "lexicon" (2000c, 52–53). He argues that terms from one lexicon may be unable to be translated into another due to a restriction on relations between natural kinds. The restriction derives from a "no-overlap principle," according to which members of one natural kind may only be members of another if one kind is contained in the other (2000d, 92; 2000h, 232). A term cannot be translated from one lexicon into another if its extension includes items belonging to distinct kinds within the rival taxonomy, since that would violate the no-overlap principle.

As may be seen from this brief survey, the claim of incommensurability does not constitute a precise or stable target. There are a number of variants of the idea, and it has undergone modification. Nevertheless, the idea has been subjected to searching critique. I will now briefly indicate the most salient lines of criticism that have been raised against it.

2. THE MAIN OBJECTIONS TO INCOMMENSURABILITY

The aim of this section is to provide an overview of the main objections that were presented against incommensurability. I will start with the semantic form of incommensurability before turning to the methodological form of the doctrine.

The main critical concern with semantic incommensurability stems from the apparent inability to compare the content of incommensurable theories. If the terms of incommensurable theories have no meaning in common, or if the language of one may not be translated into the language of the other, then no assertion of one theory may assert or deny an assertion made by the other. But if the content of theories may not be compared, it would be impossible to conduct a comparative assessment of the theories with respect to a shared body of evidence. Given this, it is not clear how it would be possible to choose between such theories on a rational basis. Indeed, it is entirely unclear why there should be any need to choose between such theories in the first place.

In fact, the thought that semantically variant theories are incomparable for content trades on a crucial ambiguity between sense and reference. As noted by Israel Scheffler (1967), comparison of content requires sameness of reference rather than sameness of sense. Two assertions whose constituent terms refer to the same things may enter into agreement or disagreement whether or not their terms have the same sense. This means that incomparability of content does not follow from the variation of meaning between theories. Only if the variation of meaning gives rise to failure of co-reference does incomparability follow. Nor is the point restricted to terms that have identical extensions. All that is required for theories to be able to enter into conflict is overlap or intersection of the extensions of their terms (cf. Martin, 1971).[6] Related remarks also apply with respect to realist concerns about progress: provided that there is shared or overlapping reference between the terms employed by successive theories, there may be progress in the sense that there is an increase in knowledge about a common domain of entities.

Apart from the issue of content comparison, the semantic incommensurability thesis also raises the prospect of communication breakdown between scientists in the context of theory-choice. Kuhn originally claimed that there is partial communication failure between proponents of competing paradigms during the revolutionary phase of a science (2012, 148). However, he came to recognize that semantic variation need not entail failure of communication (2000c, 37). The semantic variation that gives rise to incommensurability is restricted to the special vocabulary of competing paradigms. Inability to translate between semantically variant vocabularies is localized failure of translation within the context of a shared background language, including scientific terms that the theories share. Scientists from competing paradigms are able to draw upon the extensive resources of the background language, as well as the terms shared by the theories in order to communicate with each other. They may understand each other despite semantic differences between the special vocabularies of the paradigms in which they work. In addition, translation and understanding are not the same thing. Translation involves the identification of expressions of one language that are semantically equivalent to expressions of another language. Understanding an assertion involves grasping the meaning of the assertion. While an expression of one language may fail to be translatable into another, it does not follow that there must be failure to communicate. A bilingual speaker of two languages may understand an expression from one language that cannot be translated into the other. Similarly, a scientist may understand what is meant by the terms of an opposing theory even though the terms are not translatable into the vocabulary of their own theory.[7]

We may now turn to the methodological version of incommensurability. In *Structure*, Kuhn took incommensurability to be in part due to lack of shared standards between paradigms. Paradigms address different problem-sets and employ different standards of problem-solving adequacy. As a result, there are no shared standards of appraisal on the basis of which to comparatively evaluate competing paradigms. It would seem to follow that there may be no rational basis for the choice between paradigms, since there are no shared standards on which such a choice might be based. This aspect of the incommensurability thesis encountered strong resistance. Philosophers of science pointed to a range of methodological factors that might serve as an objective basis for comparative appraisal of paradigms.[8]

In response to such criticism, Kuhn claimed that he had not intended to deny the existence of stable criteria of theory-appraisal. In "Objectivity, Value Judgment and Theory Choice," he set out a list of criteria (viz., accuracy, consistency, breadth, simplicity, and fruitfulness) to which scientists appeal in the comparative appraisal of competing theories or paradigms.[9] The criteria function as values that guide theory-choice rather than as rules that dictate such choice. But while stable criteria of theory-appraisal exist, Kuhn argued that they do not constitute an algorithm of theory-choice. Individually, the criteria may be understood in different ways. Collectively, the criteria may conflict in application to competing theories (e.g., one theory may be more accurate while another is simpler). Moreover, the criteria do not admit of a fixed rank-order, though Kuhn does allow that accuracy tends to be the "most nearly decisive" factor (1977, 323).

Kuhn's view that a non-algorithmic set of theory-neutral criteria may provide a rational basis for scientific theory-choice has proven to be less controversial than his earlier apparent denial of shared inter-paradigmatic standards.[10] Despite this, one shortcoming with Kuhn's view is that he was unable to provide a satisfactory account of the normative basis for the criteria of theory choice. Some philosophers have thought that the basis must be a priori. Others (myself included) favor the development of a naturalistic account of the epistemic warrant of the norms of scientific method.[11]

These last remarks conclude my discussion of the methodological version of the incommensurability thesis in the present context. For the remainder of this chapter, I will restrict discussion specifically to the semantic version of the incommensurability thesis. My reasons for narrowing the focus of the discussion in this way are twofold. First, the primary use of the term "incommensurability" relates to the semantic form of the doctrine. This is certainly the case with respect to Feyerabend, who with one exception restricts his use of the term to semantic variance between theories. As for Kuhn, we have just seen that in his own work on the topic he came to focus exclusively on semantic incommensurability, having renounced the methodological version of the thesis. Second, in the remainder of this chapter, my aim is to bring the question of the relationship between incommensurability and the incomparability of the content of theories into sharper focus. I regard this question to be the sole issue that remains to be resolved in relation to the incommensurability thesis.

3. ON THE SO-CALLED PHENOMENON OF INCOMMENSURABILITY

Despite serious objections being leveled against the thesis of incommensurability, the doctrine continues to enjoy a modicum of support. Indeed, some authors appear to hold that incommensurability is a phenomenon the existence of which has been established. The purpose of this section is to raise doubts about the claim that there is a phenomenon of incommensurability.

In asking what the argument is for incommensurability, Moti Mizrahi notes the use made by Eric Oberheim and Paul Hoyningen-Huene of the language of discovery when they write about incommensurability. In their *Stanford Encyclopedia of Philosophy* entry on the topic, they report Kuhn as having claimed that "he discovered incommensurability as a graduate student in the mid to late 1940s while struggling with what appeared to be nonsensical passages in Aristotelian physics" (2013, section 2.2.1). He later left theoretical physics to "pursue this strange phenomenon." Mizrahi comments that: "Using the 'discovery' language in talking about Kuhn's incommensurability thesis gives the impression that incommensurability is a fact about scientific change (revolutionary change, in particular)" (2015, 362). This leads Mizrahi to wonder whether there are any compelling grounds for holding that incommensurability has been established as a fact.

In addition to the passage cited by Mizrahi, Hoyningen-Huene and Oberheim refer to incommensurability as a phenomenon in another context. In a comment on a paper of mine about the relation between semantic incommensurability and scientific realism, they write as follows:

For Feyerabend and Kuhn, incommensurability is not *based* upon anti-realist metaphysics, but rather it resulted from analysis of the historical phenomenon of incommensurability, which in turn *resulted* in doubts about realism, and increased the plausibility of some sort of neo-Kantian metaphysics. Again, the argument can be construed as an inference to the best explanation: given the phenomenon of incommensurability as apparent in the historical record, its best explanation consists in the assumption of a neo-Kantian metaphysics (Hoyningen-Huene and Oberheim 2009, 208).

Hoyningen-Huene and Oberheim see a close connection between incommensurability and the antirealism they ascribe to Feyerabend and Kuhn. Accordingly, they take the use of realist assumptions about reference in arguing against incommensurability to beg the question. In the passage just quoted, they describe incommensurability as a "historical phenomenon," which is "apparent in the historical record." In their view, Feyerabend and Kuhn arrive at an antirealist metaphysical view on the basis of an inference from the historical phenomenon of incommensurability to the best explanation of that phenomenon.

I regard this account of the relation between incommensurability and antirealism as dubious.[12] But that is not the point at issue here. What is at issue is whether it makes sense to speak as if there is a phenomenon of incommensurability. Feyerabend and Kuhn employ the term "incommensurability." They apply it to a range of different cases. They make claims about change of meaning and conceptual apparatus in theory-change. Does it follow that there is a phenomenon of incommensurability?

In the case of Feyerabend's original discussion, there are three distinct components of the claim of incommensurability. One component is the realistic interpretation of theory, according to which observational terms receive their meaning from the theory in which they are employed. The second is the claim that the term "impetus" is unable to be defined in the context of a theory that rejects laws on the basis of which the term is defined. The third is the claim that as a result of the first two points, the consequences of meaning variant theories are incomparable.

It is not immediately clear which, if any, of these three components of the claim of incommensurability might be regarded as picking out a phenomenon. The realistic interpretation of theories is a controversial philosophical claim about the nature of observational meaning that does not enjoy universal assent. The claim that the term "impetus" is unable to be defined in the context of a competing theory may seem to report a phenomenon. But this too is unclear. Feyerabend presents a detailed argument for the indefinability of "impetus," which turns on non-trivial considerations that may well be found contentious. A similar remark applies to the claim about incomparability. As we have seen, the claim that meaning variant theories have no comparable consequences has been denied on the basis of the distinction between sense and reference. In sum, it is at the very least not obvious why Feyerabend's original discussion of incommensurability should be taken as reporting the discovery of a phenomenon.

The situation may be different with Kuhn. Oberheim and Hoyningen-Huene draw attention to Kuhn's description of his encounter with Aristotle's physics, which led him to the idea of incommensurability. At first, Kuhn found Aristotle's text deeply puzzling. Aristotle was an influential figure in the history of physics. And yet, as Kuhn read him, he "appeared not only ignorant of mechanics, but a dreadfully bad physical scientist as well" (2000b, 16). But as Kuhn continued to struggle with the text, all at once it made sense: "Suddenly the fragments in my head sorted themselves out in a new way, and fell into place together" (2000b, 16). To understand Aristotle, Kuhn had to acquire the Aristotelian conceptual apparatus rather than apply the Newtonian conceptual framework with which he began.[13]

What Kuhn describes certainly appears to be a recognizable phenomenon. One is presented with a text or perhaps a spoken utterance, which one does not at first understand. It subsequently turns out that the initial failure to understand the text or utterance is due to the fact that it contains words that do not have the meanings that one took them to have. The failure to understand is due to misinterpretation of the meaning conveyed by these words. But once the words are appropriately interpreted in terms of the author's or speaker's actual meaning, it becomes possible to understand the text or utterance.

Let us grant that such episodes of comprehension following initial failure to understand constitute a recognizable phenomenon.[14] But it is simply not clear that the actual phenomenon itself is best described as a case of incommensurability. The phenomenon is the act of comprehension following the initial failure to understand. The notion of incommensurability is appealed to as an explanation of one such episode. Kuhn's explanation of why he initially failed to understand Aristotle is that he had not originally understood the meaning of Aristotle's terms. Instead, he brought his own Newtonian conceptual apparatus to bear on Aristotle's text when he first read it. That is why the text made no sense to him. He only came to understand the text once he acquired the Aristotelian concepts and interpreted the terms in the text with the meanings attached to them by Aristotle. The phenomenon is thus the act of comprehension following the initial failure to understand the text of Aristotle. The explanation of the phenomenon proposed by Kuhn is that it was due to the fact that he brought an incommensurable conceptual apparatus to bear on the text: only once appropriate conceptual apparatus was in place could the text be understood. In short, it seems to me to be a mistake to think of Kuhn's encounter with Aristotle as discovery of the phenomenon of incommensurability rather than the experience of understanding an initially incomprehensible text.

Given what I have said in relation to Feyerabend and Kuhn, I suggest that there are substantial grounds to doubt that there is a phenomenon of incommensurability whose existence has been established. As we shall see in the next section, we must think in more detail about just what is required in order for there to be a case of incommensurability.

4. INCOMPARABILITY OF CONTENT AND UNTRANSLATABILITY

I have just argued against the view that there is a phenomenon of incommensurability that was discovered by Feyerabend and Kuhn. But there is a further question that remains to be addressed. What exactly is required in order for there to be a recognized case of incommensurability?

As we saw in section 1, Feyerabend and Kuhn make a number of different claims in relation to incommensurability. Given the range of different claims that they make, it may be difficult to retrieve from their discussion a single precise criterion on the basis of which to identify cases of incommensurability. However, a claim that has occupied center stage in the literature (and my own writing) on the topic is the claim that the content of incommensurable theories is unable to be compared. If one is impressed by the referential response to incommensurability (as I am), then one will expect there to be no or very few actual cases of incommensurability, assuming the relevant criterion to be the incomparability of content.

The idea that incomparability of content is the key to incommensurability is not agreed to by all parties to the dispute. In their previously cited commentary, Hoyningen-Huene and Oberheim object to my emphasis on content comparison:

> neither Feyerabend nor Kuhn suggested that incommensurability precludes (rational) content comparison. In fact, both unequivocally and explicitly stated that incommensurable theories can be rationally compared on the basis of their empirical predictions. The threat that incommensurability poses to realism is *not* based on the impossibility of comparing the content of theories, and consequently Sankey's strategy to defuse it by showing that incommensurable theories can be compared with regard to content is off target (Hoyningen-Huene and Oberheim, 2009, 205).

In this passage, Hoyningen-Huene and Oberheim treat the ideas of rational content comparison and comparison of content as interchangeable. This runs together two things that should be kept distinct. It runs the idea of content comparison together with the idea of having a rational basis for a choice between theories. But never mind that. The important point for present purposes is that Hoyningen-Huene and Oberheim resist the claim that incommensurability entails the incomparability of the content of theories.

In the case of Feyerabend, the connection between incommensurability and incomparability of content is one that he made explicitly on a number of occasions. We have already seen his comment that incommensurable theories have no "comparable consequences, observational or otherwise" (1981b, 93). The point emerges very clearly in an exchange with Dudley Shapere.

Feyerabend quotes Shapere as saying that "sentences which do not have common meaning can neither contradict, nor not contradict, one another" (1981d, 115). Feyerabend agrees. He then suggests that in order for theories to be able to criticize each other there must be "methods which do not depend on the comparison of statements with identical constituents" (1981d, 115). The exchange with Shapere makes it clear that in Feyerabend's view the assertions of incommensurable theories are unable to enter into logical relations such as contradiction with each other. Feyerabend sometimes put the point in terms of the "absence of deductive relations" between theories (1982, 68). In contrasting his own view with Kuhn's, Feyerabend characterizes incommensurability in his sense as the "deductive disjointness" of two theories in the same domain (1982, 67–68). Incommensurable theories might be compared, he said, but "comparison by *content*, or *verisimilitude* was of course out" (1982, 68).

As we have already seen, the claim that semantically variant theories are incomparable for content is subject to an objection based on the sense-reference distinction that was originally presented by Scheffler. From the supposed fact that the terms of incommensurable theories differ in meaning, it does not follow that the content of such theories is unable to be compared. If the theories are applied to the same domain of entities, then, to the extent that the terms refer to anything, the terms will either co-refer or else overlap with respect to extension. Because of co-reference and extensional overlap, the two theories are able to enter into logical relations with respect to content. More specifically, claims made by one theory may agree or disagree with claims made by the other theory. As a result, the theories may be compared with respect to what they say about the world. Where the theories disagree with respect to a specific state of affairs, the disagreement may be adjudicated by empirical means. For example, if the theories disagree with respect to an empirical prediction, it may be possible to conduct a crucial test to decide the matter.

Given the role of reference in content comparison, it seems clear that incomparability of content is unable to serve as a criterion on the basis of which to establish the existence of cases of incommensurability. But perhaps there is another possibility. Let us now consider Kuhn's views in search of an alternative.

At a number of places, Kuhn writes as if theories may not be compared due to semantic difference between the language in which they are expressed (e.g., 2000e, 162 and 2000g, 204). And yet he often insists that incommensurability and incomparability are not the same. Here is one such passage:

> Most readers of my text have supposed that when I spoke of theories as incommensurable, I meant that they could not be compared. But "incommensurability"

is a term borrowed from mathematics, and it has no such implication there. The hypotenuse of an isosceles right triangle is incommensurable with its side, but the two can be compared to any required degree of precision. What is lacking is not comparability, but a unit of length in terms of which both can be measured directly and exactly. In applying the term "incommensurability" to theories, I had intended only to insist that there is no common language within which both could be fully expressed and which could therefore be used in a point-by-point comparison between them (2000f, 189).

In this passage, Kuhn denies that incommensurability entails incomparability. But it is not entirely clear whether the incomparability in question is incomparability of content. This is because he goes on to assert that "there is no common language" in which a "point-by-point comparison" may be made between the theories. This assertion may be read as the claim that it is not possible to directly compare the contents of the theories with respect to specific points of agreement and disagreement due to the lack of a common language in which both may be expressed.

Despite the lack of clarity about the relationship between point-by-point comparison and lack of a common language, it seems clear that on the whole Kuhn's intention was not to deny that the content of incommensurable theories may be compared. Indeed, the position that Kuhn endorsed in the latter part of his career allows significant scope for content comparison. Kuhn holds that translation failure between incommensurable theories is a localized inability to translate exactly between interdefined subsets of the terms used by theories. The theories share a significant amount of vocabulary within which it is possible for comparison to be made: "the terms that preserve their meanings across a theory change provide a sufficient basis for the discussion of differences and for comparisons relevant to theory choice" (2000c, 36). Outside the area in which vocabulary is shared, comparison may be undertaken on the basis of reference. Kuhn allows that the reference of expressions of a theory may be identified even though the expressions are unable to be translated into the theory with which it is to be compared (2000c, 40; 2000f, 190).

In light of Kuhn's denial of content incomparability, let us consider a further possible criterion for incommensurability. For Kuhn, what appears to be diagnostic for incommensurability is the inability to translate from the special vocabulary of one theory into the special vocabulary of another theory. Perhaps such inability to translate between the special languages of theories holds the key to incommensurability.

The trouble with this proposal is that nothing substantive would remain of the claim of incommensurability. It may very well be the case that it is impossible to translate from one narrowly circumscribed set of terms into another such set of terms given such severe restrictions on the semantic resources to be employed in the attempt to translate. But what follows? Given that the content of untranslatable theories may be compared, incommensurability so construed constitutes no impediment to rational choice between theories. Given that scientists are able to understand what is said by theories whose terms are untranslatable into their own, no insuperable obstacle stands in the way of full communication between the "proponents of competing paradigms." Given that the terms of mutually untranslatable theories may nonetheless refer to a shared domain of entities, it is entirely possible for science to progress in the sense that later theories yield an increase in knowledge about the same entities that earlier theories referred to. In sum, if untranslatability is the criterion for incommensurability, what remains of the claim has been so weakened that even if there are actual cases of incommensurability in the history of science they are of little or no interest. If mere inability to translate is the criterion for incommensurability, one wonders what all the fuss was about.

5. CONCLUSION

I began working on the problem of incommensurability in the early 1980s. At the time, it was still a live issue. The debate was well-advanced. But the issue was not yet fully resolved. Incommensurability was seen as a threat to the rationality of theory-choice, as well as to a realist view of the progress of science. Drawing on recent work in the philosophy of language, philosophers of science were working out the details of the referential response to incommensurability by developing a suitable account of reference for conceptual change in science. In addition, the question of translation was a topic of interest because the possibility of wholesale failure of translation seemed to give rise to a radical form of conceptual relativism due to alternative conceptual schemes.

Returning to the topic from the perspective of the contemporary scene in the philosophy of science is like visiting a battlefield from a forgotten war. The positions of the warring sides may still be made out. But the battlefield is grown over with grass. One may find evidence of the fighting that once took place, perhaps bullet marks or shell holes. But the fighting ceased long ago. The battle is a thing of the past.

The problem of incommensurability is no longer a live issue. The present chapter has taken the form of a *post-mortem* examination of a once hotly debated but now largely forgotten problem from an earlier period in the philosophy of science. Does anything remain of this dead issue? The thesis of incommensurability had a specific place at a specific point in the history of the philosophy of science. The heyday of incommensurability was the time of the great debates about theory-change that followed the initial publication of *Structure*. If there is a residual trace of the problem of incommensurability, it is perhaps in the sense that the idea that there is conceptual change in science now seems commonplace. But the much-feared consequences, such as incomparability, communication breakdown, and irrationality now all seem to have been greatly overblown.

There is, of course, vigorous ongoing debate with respect to the prospects of a scientific realist philosophy of science. There is, as well, significant interest in the related question of the nature and scope of scientific progress. At an earlier stage in discussion of the incommensurability thesis, the suggestion that there may be wholesale discontinuity of reference in the transition between semantically variant theories gave rise to doubts about a realist view of progress as build-up of truth about a shared domain of entities. But, with the demise of the incommensurability thesis, the debate about scientific realism is free to proceed in a manner that is unencumbered by the semantic concerns about wholesale referential discontinuity that were prompted by the incommensurability thesis.

NOTES

1 The distinction was introduced in Hoyningen-Huene and Sankey (2001, ix).

2 The one exception is Feyerabend's discussion in *Against Method* of the transition from the archaic world-view to that of the pre-Socratics in Greek antiquity (2010, chapter 16).

3 More specifically, Feyerabend argues that the concept of impetus is unable to be defined on the basis of the "theoretical primitives" of Newtonian physics (1981b, 65–67). Moreover, an empirical hypothesis that impetus co-occurs with momentum is also incompatible with Newtonian physics (1981b, 67).

4 For analysis of these developments, see my (1993) and (1998). For a glimpse of the final phase of Kuhn's thought found in unpublished material, see Hoyningen-Huene (2015).

5 One of Kuhn's main examples is the transition from Ptolemaic to Copernican astronomy: "Before it occurred, the sun and moon were planets, the earth was not. After it, the earth was a planet, like Mars and Jupiter; the sun was a star; and the moon was a new sort of body, a satellite" (2000b, 15).

6 In the literature on the referential response to semantic incommensurability, the issue of reference determination looms large. The question of the extent to which reference varies with theory change turns crucially on whether reference is determined by description or causal relation or some combination of the two. I set this topic to one side in the present discussion. But see my (1994) for an overview of the relevant literature.

7 For detailed discussion of this issue, see my (1991).

8 For some of the most influential treatments of this topic, see Lakatos (1970), Laudan (1977; 1984), Scheffler (1967) and Shapere (1984).

9 For further discussion of the shared criteria, see the "Postscript—1969" which Kuhn added to the second and later editions of *Structure* (2012, 184–86).

10 Less controversial, but not *un*controversial: Laudan points out that Kuhn's view about the role of values in theory-choice accords a pervasive role to subjective factors (1996, 14).

11 On Kuhn's failure to develop an adequate basis for the epistemic values, see Nola and Sankey (2007, 285–98). On the question of whether the normative basis of methodological rules is a priori or empirical, see Laudan's proposal of normative naturalism (1987) and Worrall's minimal a priorism (1999). For my attempt to set Laudan's normative naturalist meta-methodology within a scientific realist framework, see Sankey (2000).

12 Among other things, I am not persuaded that the neo-Kantian interpretation favored by Hoyningen-Huene and Oberheim is the correct interpretation of Feyerabend and Kuhn. In the present discussion, I am largely setting aside the question of the relationship between incommensurability and antirealism proposed by Hoyningen-Huene and Oberheim. This was a major focus of our earlier exchange. In addition to the previously cited paper by Hoyningen-Huene and Oberheim (2009), see Sankey (2009a; 2009b).

13 In effect, Kuhn's initial failure to understand Aristotle was due to his interpreting Aristotelian terms in a Newtonian sense. This might seem to constitute a counterexample to the claim of incommensurability, since the Aristotelian terms are translated into Newtonian vocabulary. But the translation was incorrect, as Kuhn realized when he came to understand the Aristotelian text. It is important to bear in mind that the translation failure that Kuhn takes to be involved in incommensurability is failure of exact translation. Mistranslation is not a counterexample to the claim that exact translation is impossible.

14 As described, the phenomenon appears to be a quite common one, occurring frequently in social interaction, and is by no means restricted to scientific revolution.

REFERENCES

Feyerabend, Paul. 1981a. *Realism, Rationalism and Scientific Method: Philosophical Papers, Volume 1*. Cambridge: Cambridge University Press.

Feyerabend, Paul. 1981b. "Explanation, Reduction and Empiricism." In Feyerabend 1981a: 44–96.

Feyerabend, Paul. 1981c. "On the 'Meaning' of Scientific Terms." In Feyerabend 1981a: 97–103.

Feyerabend, Paul. 1981d. "Reply to criticism: Comments on Smart, Sellars and Putnam." In Feyerabend 1981a: 104–31.

Feyerabend, Paul. 1982. *Science in a Free Society*. London: Verso.

Feyerabend, Paul. 2010. *Against Method* (4th ed.). London: Verso.

Hoyningen-Huene, Paul. 2015. "Kuhn's Development Before and after *Structure*." In *Kuhn's Structure of Scientific Revolutions—50 Years On*, edited by William J. Devlin and Alisa Bokulich. Switzerland: Springer International Publishing.

Hoyningen-Huene, Paul and Eric Oberheim. 2009. "Reference, Ontological Replacement and Neo-Kantianism: A Reply to Sankey." *Studies in History and Philosophy of Science* 40, 203–9.

Hoyningen-Huene, Paul and Howard Sankey, eds. 2001. *Incommensurability and Related Matters*. Dordrecht: Kluwer Academic Publishers.

Kuhn, Thomas S. 1977. "Objectivity, Value Judgment and Theory Choice." In *The Essential Tension*, 320–39. Chicago: University of Chicago Press.

Kuhn, Thomas S. 2000a. The Road Since Structure, edited by James Conant and John Haugeland. Chicago: University of Chicago Press.

Kuhn, Thomas S. 2000b. "What Are Scientific Revolutions?" In Kuhn 2000a: 13–32.

Kuhn, Thomas S. 2000c. "Commensurability, Comparability, Communicability." In Kuhn 2000a: 33–57.

Kuhn, Thomas S. 2000d. "The Road Since Structure." In Kuhn 2000a: 90–104.

Kuhn, Thomas S. 2000e. "Reflections on My Critics." In Kuhn 2000a: 123–75.

Kuhn, Thomas S. 2000f. "Theory Change as Structure Change: Comments on the Sneed Formalism." In Kuhn 2000a: 176–95.

Kuhn, Thomas S. 2000g. "Metaphor in Science." In Kuhn 2000a: 196–207.

Kuhn, Thomas S. 2000h. "Afterwords." In Kuhn 2000a: 224–52.

Kuhn, Thomas S. 2012. *The Structure of Scientific Revolutions* (4th ed.). Chicago: University of Chicago Press.

Lakatos, Imre. 1970. "Falsification and the Methodology of Scientific Research Programmes." In *Criticism and the Growth of Knowledge*, edited by Imre Lakatos and Alan E. Musgrave, 91–196. Cambridge: Cambridge University Press.

Laudan, Larry. 1977. *Progress and Its Problems*. Berkeley: University of California Press.

Laudan, Larry. 1984. *Science and Values*. Berkeley: University of California Press.

Laudan, Larry. 1987. "Progress or Rationality? Prospects for a Normative Naturalism." *American Philosophical Quarterly* 24, 19–31.

Laudan, Larry. 1996. *Beyond Positivism and Relativism*. Boulder: Westview Press.

Martin, Michael. 1971. "Referential Variance and Scientific Objectivity." *British Journal for the Philosophy of Science* 22, 17–26.

Mizrahi, Moti. 2015. "Kuhn's Incommensurability Thesis: What's the Argument?" *Social Epistemology* 29, 361–78.

Nola, Robert, and Howard Sankey. 2007. *Theories of Scientific Method: An Introduction*. Stocksfield: Acumen.

Oberheim, Eric and Paul Hoyningen-Huene. 2013. "The Incommensurability of Scientific Theories." *Stanford Encyclopedia of Philosophy* (2015 edition), edited by E. N. Zalta, URL = http://plato.stanford.edu/entries/incommensurability/.

Sankey, Howard. 1991. "Incommensurability, Translation and Understanding." *The Philosophical Quarterly* 41, 414–26.

Sankey, Howard. 1993. "Kuhn's Changing Concept of Incommensurability." *British Journal for the Philosophy of Science* 44, 775–91.

Sankey, Howard. 1994. *The Incommensurability Thesis*. Aldershot: Avebury.

Sankey, Howard. 1998. "Taxonomic Incommensurability." *International Studies in Philosophy of Science* 12, 7–16.

Sankey, Howard. 2000. "Methodological Pluralism, Normative Naturalism and the Realist Aim of Science." In *After Popper, Kuhn and Feyerabend: Recent Issues in Theories of Scientific Method*, edited by Robert Nola and Howard Sankey, 211–29. Dordrecht: Kluwer Academic Publishers.

Sankey, Howard. 2009a. "Scientific Realism and the Semantic Incommensurability Thesis." *Studies in History and Philosophy of Science* 40, 196–202.

Sankey, Howard. 2009b. "A Curious Disagreement: Response to Hoyningen-Huene and Oberheim." *Studies in History and Philosophy of Science* 40, 210–12.

Scheffler, Israel. 1967. *Science and Subjectivity*. Indianapolis: Bobbs-Merrill.

Shapere, Dudley. 1984. *Reason and the Search for Knowledge*. Dordrecht: Reidel.

Worrall, John. 1999. "Two Cheers for Naturalised Philosophy of Science." *Science & Education* 8, 339–61.

Part II

DEFENDING THE KUHNIAN
IMAGE OF SCIENCE

Chapter 5

The Kuhnian Straw Man

Vasso Kindi

In the present chapter, I argue that commentators who criticize Kuhn's work are most often fighting a straw man. Their target is a stereotype that is not to be found in Kuhn's texts. I will consider the charge based on the stereotype that the Kuhnian schema is not borne out by historical evidence and will argue that Kuhn's model, which is not actually what his critics take it to be, was not supposed to be based on, or accurately depict, historical facts. It was not a historical representation but a philosophical model that was used to challenge an ideal image of science. I suggest that giving a more accurate account of Kuhn's model will not only do justice to Kuhn's work but also draw attention to issues that, because of the stereotype, have remained in obscurity.[1]

1. INTRODUCTION

Kuhn's *The Structure of Scientific Revolutions*, opens with a promissory sentence: "History, if viewed as a repository for more than anecdote and chronology, could produce a decisive transformation in the image of science by which we are now possessed" (1970a, 1). Kuhn promised to transform, via his book, the image of science that scientists, philosophers, and laypersons were possessed by at the time. What was this image of science? It was the so-called textbook image of science, which was also attributed to the logical positivists.[2] What did this textbook image of science, explicitly, or implicitly involve? It involved the belief that science progresses steadily; that scientists accumulate knowledge by testing theories against nature, discarding falsified and retaining the confirmed ones; that there is uninhibited communication among scientists of all times and of all specialties, guaranteed, if needed, by recourse to a common and readily available pure observational language distinct from any theoretical ones; that scientists aim at a true depiction of the world which they approximate as they move from one theory to the next; that revolutions in science are not a problem in this steady development as they precipitate progress. Now, what was the image that Kuhn put in its place? Kuhn's account of science was succinctly and poignantly captured, in its stereotypical form, by Barry Barnes (1982, 13): "long periods of dreary conformity interrupted by brief outbreaks of irrational deviance." The long periods of dreary conformity correspond to what Kuhn labeled normal science, that is, to periods of scientific practice where scientists follow, almost dogmatically and unwaveringly, what particular exemplars dictate, while the outbreaks of irrational deviance refer to the Kuhnian revolutions with their attendant implication of incommensurability, which signifies radical discontinuity, lack of communication across the revolutionary divide and problems of comparative rational evaluation. Kuhn's talk of techniques of persuasion, to which scientists of different allegiances resort in order to convert one another when revolutions occur, exacerbated the worries about irrationality.

But, why did Kuhn want to transform the then dominant image of science? Obviously, he must have thought that there was something wrong with it. And what was wrong with it, according to the standard reading of Kuhn's work, was that the textbook image of science did not conform to the historical record and how science works. So, Kuhn had to advance a better image of science that would not only do justice to historical facts but would also be based on historical facts. It would not just be a better philosophical theory that could accommodate and be consistent with the historical facts. Rather, it would also start with the historical facts and be confirmed by historical facts. Kuhn's significant engagement with history of science prior to the writing of *Structure* and the presence of several examples from the history of science in *Structure* added credence to the view that takes Kuhn's image of science to be based on evidence drawn from history.

2. WAS KUHN'S MODEL BASED ON HISTORICAL EVIDENCE?

The standard reading credits Kuhn with a specific model of science, which is said to have been based, rather unsuccessfully, on historical evidence, and which involves the block replacement of frameworks that are separated between them by an abyss.[3] In assuming that Kuhn's model is based on historical evidence, the standard reading seems to ignore some contrary considerations:

• that Kuhn's arguments in *Structure* in defending his model are philosophical, not historical (see Kindi 2005, 504–6; 2015).

- that the logical positivists or logical empiricists who were credited, even if wrongly, with the textbook image of science never cared to have their model of science conform to historical facts. Feigl, for instance, a founding member of the Vienna Circle and a prominent logical positivist talking about the "orthodox view" of scientific theories said that "it should be stressed and not merely bashfully admitted that the rational reconstruction of theories is a highly artificial hindsight operation which has little to do with the work of the creative scientist" (1970, 13). Also, Carl Hempel, a proponent of logical empiricism in the United States insisted that the standard construal of scientific theories "was never claimed to provide a descriptive account of the actual formulation and use of theories by scientists in the ongoing process of scientific inquiry; it was intended, rather, as a schematic explication that would clearly exhibit certain logical and epistemological characteristics of scientific theories" (1970, 148). So, it would not be fair to criticize the logical positivists for failing to depict facts or to attribute to Kuhn such criticism. In fact, Kuhn never used this kind of argument against them. Hence, it is unlikely that he promised to substitute an accurate account of science for the logical positivists' inaccurate one. What is more, he believed that historical facts could be made to confirm any kind of theory: "If you have a theory you want to confirm, you *can* go and do history so it confirms it, and so forth; it's just not the thing to do" (Kuhn 2000d, 314, emphasis in the original).[4]
- that Kuhn and others like him may have initially thought that they were deriving their conception of science from historical facts but, eventually, Kuhn said he realized that this was misleading. "I and most of my coworkers thought history functioned as a source of empirical evidence. That evidence we found in historical case studies, which forced us to pay close attention to science as it really was. Now I think we overemphasized the empirical aspect of our enterprise " (Kuhn 2000b, 95). According to Kuhn, assuming a historical perspective, that is, seeing science as an ever-developing enterprise, was by itself enough to allow them to derive their model from first principles (Kuhn 2000c, 111–12).[5]

Kuhn's critics ignore all these considerations that speak against the view that Kuhn's model is empirical and go ahead to give evidence against it. So, it has been argued, on empirical grounds, that there are no Kuhnian revolutions, that scientists do communicate despite their different allegiances, that crises do not always precede revolutions, and that there are no conversion-like phenomena.[6] All these claims presuppose and target the stereotypical understanding of Kuhn's model, which, I propose, functions as a straw man. It can easily be attacked, but it is not the real thing.

Let us consider the case of revolutions. According to the stereotypical model, Kuhnian revolutions are rare, abrupt, dramatic, and transformative events, which mark sharp breakdowns that affect a large number of people. But Kuhn actually speaks of the discoveries and the inventions of theories that bring about revolutions as processes that are not "isolated events" but "extended historical episodes" (1970a, 2–4, 7, 52, 55). He believes that revolutions occur frequently and may affect "perhaps fewer than twenty-five people" (1970a, 181; cf. Kuhn 1970b, 249–50).[7] And, finally, Kuhn says that he has an "extended conception of the nature of scientific revolutions" covering different kinds (1970a, 7). So, arguably, Kuhn would not necessarily object to calling revolutions the intellectual changes in the "long seventeenth century" that Garber (2016) contrasts to what he understands as Kuhnian revolutions. According to Garber, there were no Kuhnian revolutions during this period since the Aristotelian paradigm was not challenged by a single theory but by a diverse group of alternatives. However, there is no such requirement in Kuhn's work. Kuhn himself compared revolutions that involve small paradigm changes to the "Balkan revolutions of the early twentieth century" and said that they may affect a small number of people and may be considered part of the normal course of events by outsiders (Kuhn 1970a, 92–93). He also said that he resisted pronouncing on whether a certain development in the history of science was "normal or revolutionary" unless he had done the historical work. "I usually have to answer that I do not know. [. . .] Part of the difficulty in answering is that the discrimination of normal from revolutionary episodes demands close historical study" (Kuhn 1970b, 251).[8] So, the Kuhnian revolutions found in Kuhn's texts hardly resemble the stereotype of Kuhnian revolutions.

References to techniques of persuasion and to conversion in *Structure* are also used to build the stereotypical model. They are taken to imply that, according to Kuhn, scientists change allegiance from one paradigm to the next, not by rational argument but by beguiling rhetoric and/or some mystical "thunderbolt intuition" (Daston 2016, 128). But Kuhn did not have this view. He never excluded arguments from the repertoire of scientific communication; he actually spoke of persuasive *arguments*. He explicitly said that scientists are reasonable and that "one or another *argument* will ultimately persuade many of them" (Kuhn 1970a, 158, emphasis added). What Kuhn denied was that scientists are compelled, either by logic or by experiment, to accept a particular paradigm. They may use arguments, not rhetorical tricks, to persuade, but when they advocate different paradigms, they may not share the premises of these arguments and may end up talking past each other. Thus, interlocutors, according to Kuhn, need to be persuaded about premises first, in order to proceed and accept what logically follows from them. And what will persuade them about premises are further arguments that will elaborate on the advantages, for instance, fruitfulness, of the new paradigm and the promise it holds for them (Kuhn 1970a, 199).[9]

Conversion is another battered concept in relation to Kuhn. It is usually understood as instantaneous and mystical and, therefore, rejected, but for Kuhn, it is a process that involves the entire scientific community for an extended period of time.[10] As he says in *Structure* (1970a, 152), the conversion may sometimes require a generation to be effected. He did not say that a community undergoes, in unison, a dramatic transformation overnight.[11] He did bring up the metaphor of Gestalt switch and spoke of "the 'lightning flash' that 'inundates' a previously obscure puzzle," of flashes of intuition that may come to scientists even in sleep, of "scales falling from the eyes"[12] (Kuhn 1970a, 122–23) but only to contrast all these to deliberation and interpretation. His target was the view that there are fixed and naked data that are variously interpreted via an inferential process. Against this view, Kuhn maintained that a scientist's perception is shaped by a paradigm and needs to be reeducated and reshaped when a revolution occurs (Kuhn 1970a, 112). An anomaly, according to Kuhn, is not terminated by fetching a new interpretation for the same data, an interpretation formed in an inferential and piecemeal fashion, but by transforming the experience gained with the old paradigm to a different bundle of experience (Kuhn 1970a, 123). "The process by which either the individual or the community makes the transition from constrained fall to the pendulum or from dephlogisticated air to oxygen is not one that resembles interpretation" (Kuhn 1970a, 121–22). Kuhn did not mean to turn scientific development into a mystical affair.[13] He spoke of conversion in order to stress that scientists do not move from one paradigm to the next the way one infers a conclusion in a deductive argument. Confronted with the same constellation of objects, scientists may reshuffle them and see them differently (Kuhn 1970a, 122–23). That is why Kuhn finds the transition from Newtonian to Einsteinian mechanics a very clear illustration of scientific revolutions: no new objects, no new concepts, but rather "a displacement of the conceptual network through which scientists view the world" (Kuhn 1970a, 102). Conversion is just a metaphor for a non-inferential process that would render successive paradigms commensurable by making one the logical implication of the other. Actual conversion in science is not miraculous, as the Pauline version has it, but takes time and involves arguments, education and training.

Kuhn's critics seem to be arguing as follows: Kuhn aimed to transform the image of science by which we were possessed. He tried to substitute an empirically adequate model for the empirically inadequate one found mostly in textbooks. However, he failed: his revolutionary model is not corroborated by historical evidence, or otherwise defended, so it needs to be abandoned and replaced by one that does not highlight discontinuity and dramatic change.

In opposition, I contend that Kuhn did, in fact, aim to transform the image of science by which we were possessed, but he did not offer an empirically adequate model in lieu of an empirically inadequate one. He did not base his model on historical evidence, so any criticism that simply aims to show that it clashes with historical facts cannot by itself damage the model. Historical facts can be made to conform to different philosophical accounts.

This dialectic raises, in turn, a number of questions: (1) If Kuhn did not offer to substitute a historically corroborated account of science for the one found in textbooks, what was he doing? What is the status and role of his model? (2) If the stereotypical sketch of Kuhn's model of science, that is, as a radically discontinuous and rationally unaccounted for succession of unrelated frameworks, does not find support in Kuhn's writings, as I have pointed out, was Kuhn's model closer to the traditional model of continuity and cumulative progress? (3) Why and how were Kuhn's critics led astray and built the straw man I am criticizing? (4) What is to be gained if the stereotypical model is dismantled and a more faithful account of Kuhn's work is painted? In the two remaining subchapters of this chapter, I will address these questions, in turn, starting with the first two.

3. THE STATUS AND ROLE OF KUHN'S MODEL

If Kuhn did not base his model on historical evidence, what is the status and role of his model? I tried to answer this question in Kindi (2005). I considered Kuhn's claim that his model can be derived from first principles and I argued that in *Structure*, he offered the conditions of possibility of the practice of science. These conditions involve the use of paradigms/exemplars to set rules that are followed dogmatically to shape normal science, which, in turn, defines what is normal and what is anomalous. An anomaly is eliminated when it is assimilated in a new set of rules, laid down by a new paradigm. The two sets, however, are not logically related, since what is anomalous in the first set is made "lawful" in the second. The move from one paradigm to the next constitutes a Kuhnian revolution. As Kuhn put it (1970b, 233), "[t]he existence of normal science is a corollary of the existence of revolutions [. . .] If it did not exist (or if it were non-essential, dispensable for science), then revolutions would be in jeopardy also."[14]

In Kuhn's model, revolutions presuppose the existence of normal science, which is necessary in order to provide the background of normalcy against which anomalies are to be detected and recognized for what they are—deviations from normalcy. "Novelty ordinarily emerges only for the man who, knowing *with precision* what he should expect, is able to recognize that something has gone wrong" (Kuhn 1970a, 65, emphasis in the original). Anomalies (and crises), in their turn, are bound to occur as normal science constantly generates puzzles extending the reach of paradigm and makes it more precise.[15] The greater the number of puzzles, the more articulate and complex the paradigm; the greater the precision it achieves through normal science, the more sensitive it becomes to the emergence of anomalies. Anomalies are not prompted by the world in the way falsifications are, given that normal science does not test paradigms against the world, but they surface when the apparatus provided by the paradigm to solve its puzzles (the paradigm's concepts, models, experimental devices, etc.) falls short of the needs and expectations it breeds. Some of the puzzles it generates may not be possible to solve by its own means; they may need a totally new approach. Thus, normal science, a highly conservative and dogmatic enterprise, is paradoxically the indispensable condition for novelty and radical change; it is the mechanism that makes change possible.[16] Scientific revolutions, which depend on normal science, need to occur—they do not just happen to occur—in order for science to develop and progress.[17] This is the reason chapter IX of *Structure* is entitled "The Nature and Necessity of Scientific Revolutions."

Two issues immediately emerge. The first issue is the following: (a) Is the Kuhnian model of science, with its interlocking concepts of *paradigm, normal science, anomaly, revolution, incommensurability*, a mere tautology?[18] Hanson (1965) already criticized Kuhn for wavering between a possibly false historical claim and putting forward an unfalsifiable set of definitions and asked Kuhn to disambiguate. The dilemma described by Hanson, however, does not allow for a third possibility, which will be explained later in the chapter, namely, the possibility of using Kuhn's pattern as a model. Kuhn did not propose an empirical theory nor did he pontificate from his armchair offering either a speculative developmental schema or a metaphysical truth. Kuhn's model aims to show that science is not one thing but many different things, which become possible by having scientists following rules that are set by particular paradigms. Since paradigms differ, rules will differ and scientific practice and traditions, which are shaped by these rules, will differ.[19]

The second issue is the following: (b) If scientific revolutions are necessary, why does Kuhn say in *Structure* that "cumulative acquisition of novelty is not only rare in fact but improbable in principle" (1970a, 96)? Isn't this an indication that his account is, after all, empirical and an acknowledgment that it may turn out to be wrong? If it is not empirical, why doesn't he say that continuity in science is "impossible" rather than "improbable"?[20] One possible answer is that he does not want to rule out the possibility of continuity in scientific development since there is incremental acquisition of modest novelty during normal science. But I think a more appropriate answer can be gleaned from the character of Kuhn's model. The necessity of which Kuhn speaks is conditional, not metaphysical necessity. Kuhn does not pronounce some necessary truth about the world whose denial is an impossibility. He focuses on the conditions that make science, as we know it, and its progress possible and offers a model of how these conditions work. This is not an invented model, arbitrarily imposed upon historical facts; rather, it is a model that has been informed by Kuhn's experience as a scientist and a historian. Kuhn knew firsthand what scientific training involves and learned from his historical work how variegated scientific practice is.[21] The pattern he outlined, *paradigm, normal science anomaly/crisis, revolution*, has not been inferred from history, but has been used more as an "object of comparison," which is laid against facts in order to highlight particular features of them, for example, scientific education and training on the basis of particular exemplars rather than rules, so that differences between scientific traditions are brought forward. Kuhn's model suggests a way of looking at facts. It does not tell us that facts have to conform to the proposed pattern in order to qualify as science.[22]

Does this mean that Kuhn's model is optional, that we can ignore it and use some other model to illuminate different aspects of science? Wouldn't the logical positivists' reconstructions be equally legitimate candidates? What does Kuhn's model have to recommend itself? Kuhn set himself a very specific goal: to transform the image of science by which we were possessed. This was an image that "held us captive" (PI section 115), that "[held] our mind rigidly in one position" (Wittgenstein 1960, 59) and could not let us see science differently. These are Wittgenstein's metaphors and are used by him in order to show how some preconceived ideas about meaning block us from seeing how varied language use is. So, Wittgenstein employed not only real but also imaginary examples of language use in order to combat an essentialist idea of meaning. Wittgenstein said to his student and friend Norman Malcolm (1984, 43):

> What I give is the morphology of the use of an expression. I show that it has kinds of uses of which you have not dreamed. In philosophy one feels *forced* to look at a concept in a certain way. What I do is to suggest, or even invent, other ways of looking at it. I suggest possibilities of which you had not previously thought. You thought that there was one possibility, or only two at most. But I made you think of others.

In a parallel move, Kuhn targeted an essentialist idea of science and used concrete cases from its history to show how varied scientific practice is. Unlike Wittgenstein, however, who invented fictional examples for his purposes, Kuhn had to appeal to actual cases to shake the deeply ingrained preconception that science is always one. Imagining mere possibilities in the case of science would not be as effective.

Kuhn's historical research showed him that the image of science by which we were possessed at the time he wrote *Structure* was not based on facts but was rather imposed on facts. It was a philosophical ideal that required that science be defined by a certain method and advance continuously and cumulatively. Kuhn proposed a different concept of science. This other concept was not discovered in history but emerged from the change of emphasis, from science as theory to science as practice. In the science-as-theory model, scientific theories were linguistic constructs, which acquired meaning and were assessed to be true or false through their relation to observation sentences that were supposed to link theories to experience and the world. It was an abstract model that applied equally to all scientific theories, whatever the time and place. The problems addressed in that context (confirmation, reduction, explanation) were logical problems dealing with relations between sentences, irrespective of whether these sentences represented actual statements scientists made. Now, when attention was drawn to science-as-practice, that is, to what scientists do, how they are educated and trained, questions regarding how practices are constituted and how they develop acquired prominence. Kuhn said that practices are formed around a paradigm, that is, an exemplar that is being imitated and followed.[23] By being followed, rules specific to the particular paradigm are set and scientists are trained to use language, handle instruments, conduct experiments, etc., in accordance with them. In this context, meaning does not seep upward into vessel-like concepts from the soil of experience (Feigl, 1970) but is determined by use in accordance with specific rules in each particular practice. So, practices are individuated by the particular paradigms that govern them, which means that the landscape of science becomes varied. By focusing on scientific practice as formed around paradigms (which breed normal science, give rise to anomalies and revolutions), Kuhn's model of science brought into relief a built-in mechanism for differentiation and radical change. In that sense, Kuhn's model is particularly fit for the purpose Kuhn had set, namely the transformation of the ideal image of science. This does not mean that it can serve any other purpose. For instance, if one does history of science, as Kuhn himself did, one does not have to apply or look for the Kuhnian categories in one's field of study.

Now, if Kuhn's model is not to be identified with the stereotype that requires it to depict radical discontinuity in the history of science by the block replacement of incommensurable paradigms, should we continue to see it as a revolutionary model? Have I turned Kuhn from a revolutionary to a Social Democrat, as Newton-Smith (1981, 102–24) put it? No. Kuhn's work continues to be revolutionary, not because he substituted a radically different image of science for the one we were familiar with for a long time, but because he did away with the very idea of an ideal image for science. In his model, the different ways of doing science do not fall short of the ideal and cannot be explained away by appealing to human weaknesses and idiosyncrasies. Differentiation in scientific activity, small or big, made possible by adopting different paradigms, is what makes science going. Kuhn's new image of science, which is actually a mosaic of different traditions, was not put together by generalizing from instances; it emerged once attention was drawn to what makes scientific practice possible, namely paradigms and what follows from them (normal, science, anomalies, revolutions). In accordance with Kuhn's own understanding of scientific revolutions, his revolution in the perception of science did not have to summon new facts or make new discoveries; it only needed a new perspective. Mary Hesse, in her review of *Structure* (1963, 286), captured nicely what Kuhn did:

> This is an important book. It is the kind of book one closes with the feeling that once it has been said, all that has been said is obvious, because the author has assembled from various quarters truisms which previously did not quite fit and exhibited them in a new pattern in terms of which our whole image of science is transformed.[24]

A change of perspective can bring about a completely new view of familiar things.

4. WHY WAS KUHN'S MODEL MISINTERPRETED AND WHAT IS TO BE GAINED FROM A MORE FAITHFUL READING?

I have argued earlier in this chapter that Kuhn's critics usually attack a stereotype of Kuhn's views. But why was his model misinterpreted and what will be gained if we try to redress things? Kuhn's critics misinterpreted him because they measured his model against the criteria and presuppositions of the so-called received view. In the received view, science is taken to mean scientific theories. Scientific theories are understood as sets of statements. These statements enter into relations of logical inference. In order for these inferences to go through, as in reducing one theory to another, the terms used in the statements should have a fixed, well-circumscribed, and stable meaning. The meaning of terms is acquired from the soil of experience on the one hand, and the theoretical postulates that connect them to other theoretical terms on the other. Even if there are changes in the theoretical part of meaning when revolutions occur, there always remains the observational part to guarantee continuity across theories and mutual understanding among scientists advocating different paradigms. When Kuhn, following N. R. Hanson, challenged the fixed and neutral nature of sensory experience and tied meaning to particular practices shaped by different exemplars and rules, the meaning of terms changed with the change of practice. For Kuhn's critics, who ignored practice and took meaning to include a theoretical and observational part, this meant that nothing common remained between theories. Consequently, communication across different frameworks was not any more possible and rationality was undermined as the transition from one theory to the next could not be mapped onto a deductive inference.

Kuhn's account was forced into the received view framework and found inadequate as it was taken to yield the above undesirable consequences.[25] These consequences would not follow, however, had Kuhn's account been seen outside the box of the received view. If the dimension of practice were highlighted in Kuhn's model of science, then meanings and concepts would emerge from following rules and would not be seen as vessels to be filled with observational and theoretical content.[26] Meanings and concepts would be uses of words in imitation of paradigms and would be, thus, de-hypostasized, extended in time, open-ended, and flexible. The important consequence of this shift of vision is that if concepts are seen as uses of words, then attention is drawn to what agents do rather than to the role of concepts in logical inference. This means that, in assessing the rationality of transition from one theory to the next, one need not consider arguments in the abstract, but the circumstances of word use in particular actual practices in order to review the options available to the scientists and the decisions they made. Evaluating the transition becomes a practical rather than an abstract theoretical matter.

What is to be gained, apart from hermeneutical accuracy, if we lift the stereotype which screens Kuhn's work? First of all, we would not have to address bizarre suppositions, such as, that scientists advocating incommensurable paradigms cannot meet in the same world and have lunch together,[27] or that the same individuals are cut off from and do not understand their previous selves should they change allegiance in the course of their careers as scientists. Second, we would not devote our efforts to find or establish all kinds of common elements between successive paradigms in order to vindicate continuity and rational progress in accordance with a dated philosophical ideal. Instead, we would be more inclined to turn our attention away from the theoretical contemplation, which has stiffened our thinking in a particular abstract mode, to how scientists reason and work. We would then be more prepared to recognize diversity and view not only science but also rationality or experiment, as more malleable concepts. More importantly, we would be able to reexamine all these issues anew, from a fresh perspective. For example, instead of accounting for communication in terms of common content and shared elements in the classical definition of concepts,[28] elements that we try to detect or devise, we could attend to how scientists and scientific communities employ particular words, what rules and routines they follow, what goals they pursue, what synergies they forge to be able to understand each other. One might think that this is turning philosophy into sociology or anthropology, but this is not necessarily so.[29] As in the case of Kuhn, who used historical facts to revise our conception of science, empirical considerations may be used to revisit other epistemic and, in general philosophical, issues. For instance, from this new perspective that takes concepts to be uses of words, one may want to examine how concepts are individuated, whether they are always evolving, what makes reconceptualization possible, how to understand radical change, how to differentiate between an aberrant development and an innovative approach, and so on. Or, one may also want to explore the implications of convergent thinking that is promoted during the Kuhnian normal science. Educators and psychologists usually think that convergent thinking inhibits creativity and prefer to encourage divergent thinking. Kuhn, however, considered convergent thinking a condition for creativity and innovation in science.

5. CONCLUSION

The Kuhnian straw man has been an obstacle for recognizing and appreciating the innovative character of Kuhn's work. It has distracted attention from what Kuhn has actually said and restricted the debate to worn-out arguments and the reiteration of standard topoi. If it were removed, we would be in a better position to assess what Kuhn's work has to offer. According to the present reading, Kuhn's model should not be understood as relying on historical evidence or as an unfounded schema but, rather, as a lens (or object of comparison) that highlights discontinuity and diversity in the history and the practice of science and focuses on what scientists do rather than on abstract theoretical arguments that formalize logical problems. Despite the misinterpretations of the model, some of which I have discussed, it has already succeeded in undermining the ideal image of science and in opening up new fields of study. So, I submit that before we move on to reject it as outdated, unfounded, or problematic, it would be worthwhile to first study it as what it is in order to explore whether there are aspects of it that have remained in obscurity and have not yet been taken advantage of. The usefulness of the model or its appropriateness will depend on the task we would like to undertake.

NOTES

1 I would like to thank Moti Mizrahi for his comments and suggestions that helped me to improve the chapter and both him and James H. Collier, executive editor of *Social Epistemology,* for giving me the opportunity to participate in this new dialogue on Kuhn's work.

2 Irzik (2012) shows that Kuhn's target in *Structure* was the textbook image of science and not logical positivism as it is commonly believed.

3 Galison (1997, 12) compared paradigms to "island empires."

4 Cf. Kuhn (1987, 363): "It is too easy to constrain historical evidence within a predetermined mold."

5 In Kindi (2005, 519–20), I argue, *pace* Kuhn, that assuming a historical perspective is not by itself enough to yield his model and I offer a different account of what Kuhn got from history.

6 Two of the most recent examples are the following: Lorraine Daston (2016, 128), criticizing Kuhn, says that being initiated into a new paradigm, learning how to reason with exemplars, "is a gradual process that proceeds in fits and starts, neither a thunderbolt intuition nor a conversion experience," while Daniel Garber (2009, 2016)–and quite a few other scholars–challenge the view that the scientific revolution, which is so much used to illustrate Kuhnian revolutions, is really a revolution as Kuhn described it.

7 Kuhn, in response to Ernan McMullin's distinction between shallow and deep revolutions, said (1993, 337): "Though revolutions do differ in size and difficulty, the epistemic problems they present are for me identical."

8 Cf. how Kuhn remembers his reaction when somebody from the audience in a lecture asked him whether he had found incommensurability in his historical research for the *Black Body* book: "I thought, 'Jesus! I don't know, I haven't even thought about that.' Now yes, I mean I *had* found it, and I later recognized what it was [. . .] It was a perfectly good question; I later realized how to answer it, but it just floored me at the time, and I sort of stammered around" (2000d, 314). Kuhn's reaction shows that he did not derive his concept of incommensurability from his historical work. Incommensurability was an implication of his model and he could retrospectively recognize it in history.

9 Cf. Kuhn (1970b, 234): "To say that, in matters of theory-choice, the force of logic and observation cannot in principle be compelling is neither to discard logic and observation nor to suggest that there are no good reasons for favouring one theory over another."

10 This is actually something that holds in general about conversions. William James (1997, 160), for instance, distinguishes between gradual and sudden processes of conversion. Cf., also what Lewis Rambo (2003, 214) writes about religious conversion which is usually taken to be radical, total, and sudden: "In fact, most human beings change incrementally over a period of time; even after a long process, often the change is less than a complete 180-degree transformation." Finally, the anthropologists Jean and John Comaroff (1991, 250), talking about the conversion of African peoples to Christianity, question whether the concept of conversion in "its common-sense European connotations" can grasp well "the highly variable, usually gradual, often implicit" transformations of social identities, cultural styles, and ritual practices.

11 Under the pressure of criticism, Kuhn felt, in his later writings, that he had to clarify and qualify his claims about conversion. He said that only individuals and not communities undergo Gestalt switches and using this term for communities was only metaphorical (2000a, 88).

12 This particular phrase alludes to St Paul's conversion which was sudden and transformative and was described in the same manner "and immediately there fell from his eyes as it had been scales" (Acts 9.18).

13 "It is emphatically not my view that 'adoption of a new scientific theory is an intuitive or mystical affair, a matter of psychological description rather than logical or methodological codification'" (Kuhn 1970b, 261). Kuhn is criticizing here Israel Scheffler's understanding of his work.

14 Cf. Kuhn (1970b, 249): "If there are revolutions, then there must be normal science."

15 According to Kuhn, anomalies would not appear only if theories were restricted to apply to phenomena that were already treated by the theory and presented no problems. But, that "would be the end of the research through which science may develop further" (Kuhn 1970a, 100). Kuhn says that the logical positivists tried to save theories in this way (e.g., by presenting Newtonian dynamics as a special case of Einsteinian dynamics, given certain restrictive conditions).

16 Kuhn (1970a, 97) says that revolutionary discoveries do not confront us "with mere historical accident." He also says that commitment to a paradigm is not only a prerequisite of normal science but also a prerequisite to surprises, anomalies, crises, and radical change (1970a, 100–101).

17 In his later work, Kuhn spoke more of speciation rather than of scientific revolutions comparing scientific development, with its proliferation of special disciplines, to biological evolution (Kuhn 2000b, 98; 2000c, 119). The process of speciation toward greater specialization, just like that of the scientific revolutions, is for Kuhn "inescapable, a consequence of first principles" (2000b, 98).

18 Kuhn himself raised the issue. After criticizing the logical positivists for restricting the range of applications of a theory to known phenomena so as to protect it from coming into conflict with any later theory, he says (1970a, 100–101):

> By now that point too is virtually a tautology. Without commitment to a paradigm there could be no normal science. [. . .] Besides, it is not only normal science that depends upon commitment to a paradigm. If existing theory binds the scientist only with respect to existing applications, then there can be no surprises, anomalies, or crises. But these are just the signposts that point the way to extraordinary science. If positivistic restrictions on the range of a theory's legitimate applicability are taken literally, the mechanism that tells the scientific community what problems may lead to fundamental change must cease to function.

19 Some philosophers have contrasted exemplars with rules in the Kuhnian scheme. For instance, Alexander Bird (2000, 71) writes: "It is with *rules* that Kuhn wants explicitly to contrast exemplars." Actually, Kuhn does not contrast exemplars with rules but speaks of the priority of paradigms over rules (1970a, 43–51). What he means is that scientists need to be acquainted with paradigms/exemplars first, in order to learn how to follow their lead, how to imitate them. Following any exemplar means following the rules that the exemplar sets, for example, how to model puzzles

on the exemplar and how to reach their solution (Kuhn 1970a, 189). These rules that emerge from following a paradigm and tell scientists what to do need not be explicit. The priority of paradigms/exemplars over rules that Kuhn speaks of is not only temporal but also logical. Exemplars make the specific rules that characterize a practice possible and warrant their application. Kuhn contrasts exemplars, not with rules *simpliciter*, but with those rules that are supposed to be able to dictate what is to be done independently of any concrete application. In his view, the mere expression of a rule in words "taken by itself, is virtually impotent" (Kuhn 1970a, 191). Learning to act according to rules requires prior exposure to concrete examples of use. Kuhn is opposed to the idea that a scientific methodology can be specified in the abstract, in terms of rules, and then given to scientists to follow. This approach has two faults: first, scientists would not know what to do should they be given only verbal statements of rules without concrete applications in practice, even if they understand the words the rules are expressed in. Second, a theoretical specification of rules would have to be generic, which means that all differences in application would have to be attributed to eliminable idiosyncrasies of the particular scientists. Kuhn wanted to say that diversity is an inherent characteristic of scientific practice given that scientific traditions are built around particular paradigms/exemplars instead of generic rules. For more on the relation between paradigms and rules and the influence that Wittgenstein's philosophy had on Kuhn in this respect, see Kindi (2012c).

20 I thank Moti Mizrahi for drawing my attention to these questions.

21 In Kindi (2005, 519–22) I argue that concentration on particulars, a characteristic mark of historical work, made Kuhn more sensitive to differences and helped him recognize the diversity of scientific practice.

22 "Object of comparison" is a Wittgensteinian term and although there are differences, it would be helpful to compare the role of Kuhn's model as I describe it to how Wittgenstein understands this concept. Wittgenstein set up simple language games, or used particular cases, real or invented, as models, to illuminate an issue and dissolve philosophical confusion. We should see a model, he said, "as what it is, as an object of comparison—as a sort of yardstick; not as a preconception to which reality *must* correspond. (The dogmatism into which we fall so easily in doing philosophy.)" (PI, section 131). What Wittgenstein meant was that we should not, as philosophers, be dogmatic and demand that reality conforms to the specifications set by the model of the philosopher as if this was the only appropriate way to look at things. We should use our models to present a way of conceiving things. Oskari Kuusela (2008, 125) explains: "an object of comparison is not used to make an empirical statement about any particular objects in the sense of being valid of only those objects, though perhaps inductively generalizable. Neither is the model used as a basis for a thesis that states that all objects falling under a concept *must* be." According to Kuusela, the necessity expressed by a model characterizes the model and not the objects of the investigation. In an earlier version of the PI section 131, Wittgenstein had, in parentheses, the sentence: "I am thinking of Spengler's mode of examination" (cited in Kuusela 2008, 126; cf. Wittgenstein 1998, 21). Oswald Spengler had proposed a certain organic model of cultural growth and decay in historical development and Wittgenstein criticized him for thinking that history must fit his model. As Northrop Frye (1974, 9)

put it, Spengler was "certain that history will do exactly what he says." If Kuhn is read as offering a developmental schema for the history of science, as Spengler did for cultures, then Wittgenstein's criticism of Spengler would, arguably, also apply to him. But, according to my reading, Kuhn was not doing that. His schema is similar to Wittgenstein's "objects of comparison." The difference with Wittgenstein is that Kuhn does not present his model as one of many. He has confidence that it captures crucial characteristics of the practice of science as we know it and, because of that, he thinks it is particularly effective in carrying out the task of transforming the dominant at the time image of science.

23 This thought was a major breakthrough for Kuhn in his effort to build his model. He wanted to account for the consensus among scientists but, being "enough of a historian," he knew that he could not attribute it to an agreement regarding specific definitions, rules, or axioms. "And that was the crucial point at which the idea of the paradigm as model entered. Once that was in place, and that was quite late in the year, the book sort of wrote itself" (Kuhn 2000d, 296).

24 Kuhn's approach as described by Hesse (1963) can be compared to Wittgenstein's method of assembling reminders (PI, section 127). Philosophers, according to Wittgenstein, do not need to "hunt out new facts," nor should they seek to learn anything new by their investigations (PI, section 89). The philosophical problems, Wittgenstein said, "are, of course, not empirical problems; but they are solved through an insight into the workings of our language [. . .] The problems are solved not by coming up with new discoveries, but by assembling what we have long been familiar with" (PI, section 109).

25 Cf. Kindi and Arabatzis (2012, 3) where we claim that *Structure*'s philosophical reception was shaped by the standards of a philosophy which was itself targeted by *Structure*.

26 Kuhn's emphasis on practice in relation to science has had extensive influence in the social studies of science. But, in this context, the practice of science is studied empirically with little concern for the philosophical implications of this idea.

27 Hempel (1980, 197): "How can [adherents of different paradigms] ever have lunch together and discuss each other's views?"

28 In Kindi (2012b), I discuss the classical view of concepts in opposition to the Wittgensteinian account, which, I think, has influenced Kuhn.

29 I have discussed issues that pertain to the relation between historical and, in general, empirical studies on the one hand, and philosophy on the other, in Kindi (2012a; 2014; 2016).

REFERENCES

Barnes, Barry. 1982. *T.S. Kuhn and Social Science.* New York: Columbia University Press.

Bird, Alexander. 2000. *Thomas Kuhn.* Chesham: Acumen.

Comaroff, Jean, and John L. Comaroff. 1991. *Of Revelation and Revolution,* Vol. I. Chicago: The University of Chicago Press.

Daston, Lorraine. 2016. "History of Science Without *Structure*." In *Kuhn's Structure of Scientific Revolutions at Fifty*, edited by Robert J. Richards and Lorraine Daston, 115–32. Chicago: University of Chicago Press.

Feigl, Herbert. 1970. "The 'Orthodox' View of Theories: Remarks in Defense as well as Critique." In *Analyses of Theories and Methods of Physics and Psychology*. Minnesota Studies in the Philosophy of Science Series, vol. IV, edited by Michael Radner and Stephen Winokur, 3–16. Minneapolis: University of Minnesota Press.

Frye, Northrop. 1974. "The Decline of the West by Oswald Spengler." *Daedalus*, 103:1, 1–13.

Galison, Peter. 1997. *Image and Logic. A Material Culture of Microphysics*. Chicago: University of Chicago Press.

Garber, Daniel. 2009. "Galileo, Newton and All That: If It Wasn't a Revolution What Was It?" *Circumscribere* 7, 9–18.

Garber, D. 2016. "Why the Scientific Revolution Wasn't a Scientific Revolution and Why It Matters." In *Kuhn's Structure of Scientific Revolutions at Fifty*, Robert J. Richards & Lorraine Daston (eds.), 133–48. Chicago: University of Chicago Press.

Hanson, Norwood, R. 1965. "A Note on Kuhn's Method." *Dialogue 4, 371*–75.

Hempel, Carl. 1970. "On the 'standard' conception of scientific theories." In *Minnesota Studies in the Philosophy of Science Vol. IV,* edited by M. Radner and S. Winokur, 142–63. Minneapolis: University of Minnesota Press.

Hempel, Carl. 1980. "Comments on Goodman's *Ways of Worldmaking*." *Synthese* 45:193–99.

Hesse, Mary. 1963. "Review of *The Structure of Scientific Revolutions*." *Isis*, 54, 2, 286–87.

James, William. 1997. *The Varieties of Religious Experience*. New York: Touchstone

Irzik, Gürol. 2012. "Kuhn and Logical Positivism: Gaps, Silences, and Tactics of SSR." In *Kuhn's The Structure of Scientific Revolutions Revisited*, edited by Vasso Kindi and Theodore Arabatzis, 15–40. London: Routledge.

Kindi, Vasso. 2005. "The Relation of History of Science to Philosophy of Science in *The Structure of Scientific Revolutions* and Kuhn's Later Philosophical Work." *Perspectives on Science*, 13, 4, 495–530.

Kindi, Vasso, and Theodore Arabatzis. 2012. "Introduction." In *Kuhn's The Structure of Scientific Revolutions Revisited*, edited by Vasso Kindi and Theodore Arabatzis, 1–12. London: Routledge.

Kindi, Vasso 2012a. "The *Structure*'s Legacy: Not from Philosophy to Description." *Topoi* 32:1, 81–89.

Kindi, Vasso 2012b. "Concept as Vessel and Concept as Rule." In *Scientific Concepts and Investigative Practice*, edited by Uljana Feest and Friedrich Steinle, 23–46. Berlin: De Gruyter.

Kindi, Vasso 2012c. "Kuhn's Paradigms." In *Kuhn's The Structure of Scientific Revolutions Revisited*, edited by Vasso Kindi and Theodore Arabatzis, 91–111. London: Routledge.

Kindi, Vasso. 2014. "Taking a Look at History." *Journal of the Philosophy of History* 8, 96–117.

Kindi, Vasso. 2015. "The Role of Evidence in Judging Kuhn's Model: On the Mizrahi, Patton, Marcum Exchange." *Social Epistemology Review and Reply Collective* 4:11, 25–33.

Kindi, Vasso. 2016. "Collingwood, Wittgenstein, Strawson: Philosophy and Description." *Collingwood and British Idealism Studies*, 22:1, 19–43.

Kuhn, Thomas, S. 1962/1970a. *The Structure of Scientific Revolutions*. Chicago: The University of Chicago Press.

Kuhn, Thomas, S. 1970b. "Reflections on My Critics." In *Criticism and the Growth of Knowledge*, edited by Imre Lakatos and Alan Musgrave, 231–78. Cambridge: Cambridge University Press.

Kuhn, Thomas, S. 1978/1987. *Black Body Theory and the Quantum Discontinuity, 1894–1912*. Chicago: The University of Chicago Press.

Kuhn, Thomas, S. 1993. "Afterwards." In *World Changes*, edited by Paul Horwich, 311–41. Cambridge, MA: MIT Press.

Kuhn, Thomas, S. 2000a. "Possible Worlds in History of Science." In *The Road since Structure*, edited by James Conant and John Haugeland, 58–89. Chicago: The University of Chicago Press.

Kuhn, Thomas, S. 2000b. "The Road since *Structure*." In *The Road since Structure*, edited by James Conant and John Haugeland, 90–104. Chicago: The University of Chicago Press.

Kuhn, Thomas, S. 2000c. "The Trouble with the Historical Philosophy of Science." In *The Road since Structure*, edited by James Conant and John Haugeland, 105–20. Chicago: The University of Chicago Press.

Kuhn, Thomas, S. 2000d. "A Discussion with Thomas Kuhn." In *The Road since Structure*, edited by James Conant and John Haugeland, 255–323. Chicago: The University of Chicago Press.

Kuusela, Oskari. 2008. *The Struggle Against Dogmatism. Wittgenstein and the Concept of Philosophy*. Cambridge, MA: Harvard University Press.

Malcolm, Norman. 1984. *Ludwig Wittgenstein: A Memoir*. Oxford: Oxford University Press.

Newton-Smith, W. H. 1981. *The Rationality of Science*. Boston: Routledge & Kegan Paul.

Rambo, L. R. 2003 "Anthropology and the Study of Conversion." In *The Anthropology of Religious Conversion*, edited by Andrew Buckser and Stephen D. Glazier, 211–22. Lanham, MD: Rowman and Littlefield Publishers.

Wittgenstein, Ludwig. 1958/1960. *The Blue and Brown Books*. New York: Harper and Row Publishers.

Wittgenstein, Ludwig. 1980/1998. *Culture and Value*, edited by Georg Henrik von Wright and Heikki Nyman, translated by Peter Winch. Oxford: Blackwell.

Wittgenstein, Ludwig. 1951/2009. *Philosophical Investigations*. Translated by G. E. M. Anscombe, P. M. S. Hacker and J. Schulte. Oxford: Blackwell. Abbreviated as PI.

Chapter 6

Kuhn, Pedagogy, and Practice

A Local Reading of Structure[1]

Lydia Patton

1. INTRODUCTION

Moti Mizrahi has argued that Thomas Kuhn does not have a good argument for the incommensurability of successive scientific paradigms. With Rouse, Andersen, and others, I defend a view on which Kuhn primarily was trying to explain scientific practice in *Structure*. Kuhn, like Hilary Putnam, incorporated sociological and psychological methods into his history of science. On Kuhn's account, the education and initiation of scientists into a research tradition is a key element in scientific training and in his explanation of incommensurability between research paradigms. The first part of this paper will explain and defend my reading of Kuhn. The second part will probe the extent to which Kuhn's account can be supported, and the extent to which it rests on shaky premises. That investigation will center on Moti Mizrahi's project, which aims to transform the Kuhnian account of science and of its history. While I do defend a modified kind of incommensurability, I agree that the strongest version of Kuhn's account is steadfastly local and focused on the practice of science.

2. IMAGES OF SCIENCE

Science may move through time by gathering results and facts, and developing increasingly sophisticated methods for dealing with them. Scientists may become better over time at describing, understanding, and explaining the phenomena they encounter. Those phenomena are real and publicly available, and successive scientific theories hone in on increasingly accurate analyses and predictions of their properties and behavior.

Thomas Kuhn's *The Structure of Scientific Revolutions* unsettles this image of science. Kuhn begins by asserting that the behavior of *historians* of science cannot be explained on a cumulative picture of science. Historians may attempt to reconstruct science's past on the assumption that science is a cumulative, continuous practice. But that assumption quickly becomes a hypothesis, which is falsified as the historians dig deeper.

Kuhn cites comprehensive historical studies of research traditions in optics, electromagnetism, and related fields, by Alexandre Koyré and numerous others, which reveal breaks and discontinuities in the description of scientific practice revealed by its history.[2] Those discontinuities are found in the behavior and practice of the scientists who were working at the time. Kuhn thus defends two hypotheses: the first about scientific practice, and the second about the behavior of historians of that practice. In both cases, I will argue, his aim is to describe, and then to explain, human behavior.

Kuhn constructs a framework with which to explain scientific practice in the past, intended to extend to scientific practice generally. According to that framework, most science is essentially conservative, in the sense that it preserves the achievements that are taken as models for scientific work. Scientists are trained in a way of approaching problems and puzzles that ossifies great scientific achievements, turning them into the skeleton of a research tradition. Flesh is put on the bones by laboring researchers.

Scientists in the wake of a great scientific achievement are trained to rebuild the skeleton of that achievement, before they begin to fill out that skeleton as mature researchers. Examples of scientific achievements on the scale Kuhn analyzes are Lavoisier's *Chemistry*, Newton's *Principia*, and Franklin's *Electricity*. Thousands of scientists of the past have been trained to reproduce the achievements of these books, so that the strategies found therein become working scientific instruments.

According to Kuhn's original definition in the second chapter of *Structure*, a paradigm is a scientific achievement that becomes a textbook.[3] Books like *Principia*, *Chemistry*, and *Electricity* lay down firm results, but are also open-ended, so that scientists can find intriguing problems to solve using those results and achievements as a springboard. A young scientist doesn't just learn established theory by being taught from Newton's *Principia*. That scientist learns how to become an active researcher, how to approach problems, how to think about and represent the phenomena under investigation, and how to use instruments to conduct experiments.

What many researchers miss in Kuhn's definition are the link to *scientific practice* and the link to *teaching science*.[4] A "paradigm" often is treated as if it is identical to a "background theory," and then paradigms are fed into the machine of confirmation, testing, and assessment of theories. But Kuhn wanted paradigms to be about future research, and to be linked to scientific practice.[5] When a paradigm guides scientific practice in the right way, it makes scientific research and progress possible, and it goes beyond accepted theory.

Scientists who learn how to use a scientific achievement as a playbook are doing "normal science." Kuhn's description of normal science as conservative and dogmatic is notorious.[6] Critics complained right away that normal science only extends a paradigm that is taken as given, and does not subject that paradigm to rigorous testing. On Popper's view, for instance, if a research practice does not include at least the possibility of falsifying the theory in the background or the hypotheses under investigation, it is not scientific.

However, paradigms, in the 1962 *Structure*, are neither theories nor hypotheses. They are model scientific achievements. Strictly speaking, a theory cannot be a paradigm in the sense discussed earlier. Theories can be tested precisely because they contain sets of assertions that have fixed, knowable truth-values. In this sense, theories are closed. If an assertion of a theory does not have a truth-value that can be assessed in principle, or if that assertion cannot be proven, the assertion is not a result of the theory. Thus, if Newton's *Principia* contained only a theory or theories, it could not be a paradigm. *Principia* is a paradigm because, along with the stable results it states, the work implicitly expresses a new way of doing science. Newton's achievement results from a novel orientation to scientific practice, an orientation that can be learned and can be the source of a tradition of research.

In 1959, Kuhn described the interplay between tradition and innovation as the "essential tension" without which science cannot operate. The essential tension is the source of the theoretical posit at issue in a recent exchange between Moti Mizrahi (2015a, 2015b, 2015c), James Marcum (2015a), Vasso Kindi (2015), and myself (Patton 2015b): the incommensurability of successive paradigms.

3. INCOMMENSURABILITY

On Kuhn's account, the only reason for a working scientist to question a paradigm comes when there is a persistent anomaly: when experiments set up using the paradigm begin to fail, when the paradigm fails to solve new problems, and so on. The first task of a working scientist is to resolve anomalies within the paradigm. If that fails, then the scientific community must—reluctantly—find a way to conceive of a new paradigm to resolve persistent anomalies.

One puzzling feature of Kuhn's "essential tension" is the dual role of the architects of a paradigm. On Kuhn's account, paradigm architects are treated with reverence by the generations who follow them. The main lines of their research programs are traced and retraced by generations of scientists. We might think of graduate students learning to reproduce Fourier transforms, or learning to use Hamiltonians in simplified cases, as monks copying texts in a scriptorium. Even more profoundly, scientists being trained in a paradigm learn how to *think about* and how to *conceive of* scientific phenomena through this training, including how to describe target phenomena so they are tractable by scientific methods (see Andersen 2000).

On the other hand, it is only a matter of time before any given architect of a paradigm loses his or her place in the pantheon. No program of normal science is immune to persistent anomaly. It is an axiom of Kuhn's account[7] that no paradigm can deal with all the phenomena. The works of Aristotle were treated by the scholastics with reverence. But once prominent cases were made that the effectiveness of the Aristotelian paradigm was limited, Aristotle had to be displaced in a revolution. Kuhn's account turns Aristotle, Newton, and Einstein into warring Greek gods.

Kuhn's picture of the history of science replaces serene continuity with political upheaval. A paradigm is constructed only when persistent anomaly dogs the old paradigm: when there is a recognized crisis. The only way to deal with crisis is to change the fundamental approach to the problem. Another approach must be achieved that changes how scientists interact with the phenomena in practice.

Scientific revolutions are best explained as practical decisions, from either inside or outside an existing paradigm (Patton 2015b). Scientists will change their paradigm only when forced to do so, according to Kuhn. When they do, it is because the existing paradigm no longer works. Scientists have been trained to approach the phenomena in a certain structured and artificial way, a way not limited to accepting the claims of the background theory. The approach may involve being trained into practical ways of modeling the phenomena and ways of setting up experiments, for instance, practical know-how that Michael Polanyi calls "tacit knowledge."[8]

Scientists are trained as well in what Ludwik Fleck calls "vademecum science" and Kuhn calls "textbook science," in contrast with "journal science."[9] Textbook science is the image of science sketched earlier in the text, in which students are initiated into their scientific community's way of approaching the phenomena and of solving problems. As Hoyningen-Huene puts it, this training even gives students "access to the (region of the) phenomenal world relevant to the work of his or her community."[10]

Once students have been trained, they are set loose on the world as researchers, and, again, the phenomena are not always tractable by a given paradigm. Researchers will find anomalies and misfits in practice, most of which can be resolved. But if a crisis occurs, and then a revolution, then there must be a "paradigm shift." Paradigm shifts, for Kuhn, involve changes to the conceptual and semantic categories in play. A new paradigm may even mean that scientists educated in the former tradition may need to "reeducate" themselves to perceive the world differently (Kuhn 2012/1962, 112).[11]

The earlier Kuhn of *Structure* presented incommensurability as a kind of practical impossibility. Imagine a scientist, Alessandra, has been trained in the fluid theory of electricity.[12] Alessandra knows how to work with an early kind of battery called a Leyden jar, a glass jar filled with a fluid (acid) with immersed wires. As Alessandra has been trained to conceive of it, the fluid is necessary to produce a current across the wires. Now imagine Alessandra is in her lab, looking for a battery to use in her experiment. There is a stack of dry cell batteries in the corner, which are composed of a chemical paste and metal contacts. But she looks past them, and says, "There aren't any batteries here." To her, a dry cell cannot be a battery, because there is no fluid involved.

For Alessandra to be able to use the dry cells to generate a current for her experiment, or to draw any conclusions from that experiment, she will need to change her way of working with electricity. That need for change to a scientist's way of working is a practical result of incommensurability: former paradigms clash with practices, claims, or structures that emerge under new paradigms. Scientists can conduct certain experiments or prove certain results within one paradigm and not another. The paradigms can be compared to each other, but there is no common measure that reduces one to the other.[13] Certainly, it is true that a scientist can change her way of thinking and her way of working. But she cannot work exclusively within an unrevised fluid electricity paradigm and work with dry cells successfully at the same time.

4. THE PROBLEM WITH INCOMMENSURABILITY

The continuous explanation of science that Kuhn rejects has two attractive, connected features: continuity and realism. These features are central to contemporary realist accounts. Since scientists are referring to the same things, their descriptions and explanations have a secure basis for comparison over time. The history of science has a foundation of reference to real, publicly available phenomena. The behavior of scientists can be explained in these terms as well. Scientists make the inferences they do, and construct the theories they do, because their experiments and investigations put them in causal contact with objects and systems with stable properties.[14] The statements of scientific theories are intended as descriptions of those properties, and, when they fall short, they are corrected when the evidence is updated.[15]

The advantages of continuity and realism are palpable when there are rival approaches to the same phenomena. It is straightforward to say that theories compete to be the best descriptions and explanations of interesting phenomena. Moreover, we can explain scientists' behavior by rationally reconstructing[16] how they react to novel evidence, counterexamples, the development of new analytic tools, and the like. Such a reconstruction may be easier if we think of scientists using rival approaches as trying to measure the same things using different yardsticks.

Ever since Kuhn published *Structure*, there have been criticisms of the notion of incommensurability, and of Kuhn's related claim that paradigm shifts are not rationally reconstructible.[17] Fundamentally, one might deny Kuhn's thesis of the "priority of paradigms." We might paraphrase the priority thesis as the assertion that a scientist's way of working with the phenomena, "the set of results provable, puzzles solvable, and propositions cogently formulable" by a scientist, depends on the paradigm under which she is working (Patton 2015b, 57).

Denying the priority thesis has a number of apparently salutary results, related to the advantages of convergent realism. Paradigm shifts become rationally reconstructible, at least in principle, because there is a perspective from outside any given paradigm from which to evaluate competing paradigms. The incommensurability of successive paradigms is undermined as well, for the same reason. It is no longer impossible in principle to find a common measure with which to evaluate competing paradigms. The threat of "Kuhn loss," in which results achieved in one paradigm are not recoverable in a successive paradigm, no longer looms over science.

By these means, we regain an image of science and its practice that preserves a robust continuity consistent with convergent or, as Mizrahi (2013) has defended recently, relative realism.[18] The continuity involved could be continuity of empirical results and practices, or of equations and structural relationships. An essential assertion of many of Kuhn's critics is that there will always be a common measure between any two paradigms, according to which results and assertions of one can be recovered in the next.[19] Scientists need not find themselves blinkered by their training into seeing the world as it is structured by an artificial approach to problems.

Denying the thesis of the priority of paradigms also removes a barrier Kuhn had placed in the way of making the following assertions:

- In principle, there could be an epistemic standard that governs scientific practice and theory, in the past and in the present alike.
- The evidence for scientific claims is publicly available, inferences from such claims are based on fundamental rational or logical principles, and thus scientific research does not require initiation into a scientific élite.
- Scientists work in a common world and with publicly available phenomena.

It is a historical irony that these or similar assertions were characteristic of the Unity of Science movement, in whose *International Encyclopedia of Unified Science Structure* first appeared (Carnap, Morris, and Neurath 1970). As Wray (2016) notes, in a 1963 letter to Kuhn, Marjorie Grene "expresses surprise at where the book is published. 'It seems a bit of a joke that it should appear in the Unity of Science Series, of all places.' "[20] What's funny is that many of Kuhn's conclusions undermine the tenets of the Unity of Science program, and vice versa.

For instance, an account of scientific observation via Otto Neurath's protocol propositions (*Protokollsätze*) that operationalize publicly available observations seems to be ruled out by Kuhn. Kuhn's scientific observation is theory-laden and highly structured.[21] Conversely, Neurath's program of Protokollsätze supports an account of scientific observation that would put the brakes on *Structure* from the beginning. As Massimi (2015) has argued in detail, Kuhn argues at least for the semantic mind-dependence of scientific phenomena.[22]

Thus, we might think that Kuhn is arguing that researchers in rival paradigms are unable to understand each other in principle, not just disinclined to do so. They "work in different worlds," speak different languages, and are unable to cross the gulf of understanding.

5. PARADIGM INADEQUACY AND SUPER-PARADIGMS

In fact, we can say more. The following is an often unacknowledged premise of *Structure*:

> *Paradigm Inadequacy*: Not all phenomena are accessible, and not all scientific results are provable, from any single paradigm.

The careful work of Brorson and Andersen (2001) and of Hoyningen-Huene (1993) provides detail to the account according to which Kuhnian researchers working within a paradigm gain access to phenomena only from within a given paradigm, where that paradigm involves training and an artificial, structured approach to investigation, experiment, and inference.[23] Kuhn draws not only on the analysis of "vademecum" or "textbook" science from Fleck but also on the work of James Bryant Conant and the Harvard science studies curriculum (Wray 2016), on the notion of "tacit knowledge" from Polanyi (Timmins 2013), and on related notions from Toulmin and Foucault.

None of this work provides sufficient evidence for Kuhn's thesis of paradigm inadequacy. It can be true that researchers gain access to phenomena only from within a given paradigm but false that no *single* paradigm provides access to all the phenomena. Without the premise of paradigm inadequacy, though, many of Kuhn's assertions lose their force. Scientific revolutions, and the resulting incommensurability, would be temporary phenomena: mere inconveniences along the way to developing an even more powerful paradigm that dominates the old and the new approaches.

Kuhn might observe that no single paradigm *ever has* provided access to all scientific phenomena or to a way of solving all extant scientific puzzles. Kuhn could respond quite simply that his account is intended as a description and an explanation of the scientific past and present, which makes sense of scientific practice. The history of scientific practice is a history of warring paradigms, not of peaceful agreement.

Technically, paradigm inadequacy is a falsifiable claim. If someone were to develop a semantic and practical magic bullet, a scientific framework from which all scientific results are recoverable and which is perfectly transparent to all forms of scientific research both formal and experimental, that would falsify it. Scientists trained in a super-paradigm would be able to do research in any domain using the paradigm as a guide for their research; they would find that results in that domain immediately apply to phenomena in related domains; and the super-paradigm would show them how to use results in one domain to solve related puzzles elsewhere.[24]

Facts about the practice and development of science require a super-paradigm, a way of solving problems that works for every science, to pursue this way of falsifying the claim of paradigm inadequacy. As David Hilbert has emphasized, questions within physics are suggested by progress in mathematics, and vice versa. It is well-known that approaches to problems in chemistry affect practice in biology, and vice versa. And so on. Without a super-paradigm, there is always the possibility that a paradigm in a single given domain will fail when that domain is extended or drawn differently, to include problems and approaches within another science.

Fortunately for them, Kuhn's critics do not need to achieve a super-paradigm. They can make one of two moves instead:[25]

1. Question the evidence for paradigm inadequacy, even in the history of science.
2. Argue against an assumption behind the premise: that a "single" paradigm must be simple and not composite. If there always will be a bridge between successful paradigms, so that results in one can be assessed and rederived from the perspective of another, that is the practical equivalent of a single super-paradigm.

Many of Kuhn's critics (Lakatos, Friedman, etc.) have taken option 2. Mizrahi (2015a and forthcoming) takes both options, unifying the fronts against

Kuhn. He argues, in a forthcoming paper, that it is not true that the history of science is a "graveyard" of past theories, as Kuhn, Laudan, and others have asserted. While not formulated explicitly in these terms, this is option 1: to deny that the history of science is a history of warring paradigms, and to assert that science displays an underlying continuity.

Stephen Toulmin has argued that Kuhn's history of science is lacking, on different but related grounds:

> with experience, it has become clear to political historians that nothing is achieved by saying "and then there was a revolution," as though that exempted one from the need to give any historical analysis of a more explicit kind. To do only that is not to perform the historian's proper intellectual task, but to shirk it (Toulmin 1967, 84).

Toulmin objects to Kuhn's depiction of the history of science as displaying, not continuity, but radical breaks. Telling the history of science requires an adequate and comprehensive description and explanation of events. Appealing to a "revolution" is a trick. You can explain what happens up to the time of the revolution (at a time t_1, say) and what happens after t_1. But you do not consider yourself responsible for explaining the break at t_1, or the relationships between the "revolution" at t_1 and what happens before and after it. That, says Toulmin, is bad history. Toulmin's objection undermines the evidential base for Kuhn's assertion that the history of science is punctuated with breaks and revolutions. For Toulmin, the "evidence" for this assertion comes from failures of historical rigor: the historian finds a gap she cannot explain, and hypothesizes that this gap is *in the history* instead of *in her explanation* of the history.

Mizrahi's and Toulmin's objections point to a unifying theme in the criticisms of Kuhn, a way to unify (1) and (2). Understanding the history of science requires not just describing events but also explaining why they happened. That requires a standard, a common measure, that spans the history of science.

One popular strategy for providing such a measure is to argue that scientists in successive paradigms are *referring to* the same things, which amounts to a denial of Kuhn's thesis of "taxonomic incommensurability."[26] For instance, Leplin and others have defended "methodological realism," which includes the claim that scientific practice makes sense only if scientists understand themselves to be working with real things that in principle are accessible to other scientists.[27]

I do not believe that there is a global argument for the essential or universal incommensurability of rival scientific theories (see Patton 2015b, 2012). To that extent, I believe that Mizrahi (2015a) and I are in agreement. But, as should be clear by now, I do not agree that a paradigm is restricted to a theory that consists of assertions with truth values that depend on the existence and properties of the referents of their terms. Instead, I think a Kuhnian paradigm is a guide to practice within a scientific community. The interesting facet of Kuhnian incommensurability is thus that it is *practical* and *local* and, as I will conclude, can contribute to the understanding of particular historical events.

6. SCIENTIFIC PRACTICE AND LOCAL EXPLANATION

Kuhn shifted from his earlier emphasis on practical incommensurability, in *Structure*, to an emphasis on taxonomic incommensurability in later work.[28] In *Structure*, however, Kuhn places the main emphasis on practice, and thus on the account that I sketched in the opening sections of this chapter. On page 103, where he introduces the term "incommensurable," Kuhn writes:[29]

> paradigms differ in more than substance, for they are directed not only to nature but also back upon the science that produced them. They are the source of the methods, problem-field, and standards of solution accepted by any mature scientific community at any given time. As a result, the reception of a new paradigm often necessitates a redefinition of the corresponding science. Some old problems may be relegated to another science or declared entirely "unscientific." Others that were previously non-existent or trivial may, with a new paradigm, become the very archetypes of significant scientific achievement. And as the problems change, so, often, does the standard that distinguishes a real scientific solution from a mere metaphysical speculation, word game, or mathematical play (Kuhn 2012/1962, 103).

The last three examples are suggestive. What counts as a scientific problem? What counts as a solution? What is trivial, and what is interesting? What is a clever solution to a merely intellectual puzzle, and what is a substantial contribution, a real scientific achievement?

Kuhn refers to these aspects of local, communal scientific *practice* when he first defines incommensurability. Shifting paradigms can result in shifting community standards: what was an uninteresting problem can become interesting, and what was scientific can be seen as *un*-scientific. Within some traditions of natural philosophy, theology is continuous with physics, because the laws of nature are willed by God. Descartes thought his account of God was a necessary support for his account of the laws of nature, which, in turn, is central to his natural philosophy. He would not have divided the two pursuits, either. But someone in the contemporary context who studies theology is not considered to be pursuing the science of physics as it is done at MIT.[30] That is a sociological fact about the way we organize scientific pursuits, the way disciplines are divided, and the way we divide up problems among researchers. But it is also a practical way that Cartesian natural philosophy and current research paradigms differ.

Contemporary scientists do talk about God or the divine, but arguably they do not count results that are *only* statements about the existence or properties of God as the sole basis for demonstrations of results within physics. There are ways to interpret cosmology, for instance, which appeal to the divine. But it is likely that a contemporary physicist who submitted a proof to *Physical Review Letters* that depended *only* on statements about the nature and attributes of God would have that paper rejected as outside the scope of the journal. Researchers in Cartesian and Newtonian natural philosophy presented such proofs to the scientific community, and they were accepted as proofs within natural philosophy.

Note that Kuhn refers this practical result of paradigm shifts first and foremost to scientific practice, to choices within the scientific community. There is nothing necessary, much less logically necessary, about it. When Laplace was asked about the place of God in his system of physics, he (allegedly) replied, "I have no need of that hypothesis." Laplace developed a scientific achievement, a system, and an approach that broke with the tradition of natural philosophy in turning away from theological concerns. Laplace's system differs from Descartes' system in practice, in its results, and in its standards of explanation. For instance, Laplace's laws of nature are not concerned with or founded on the divine essence or will, while Descartes' are.[31]

We cannot understand Laplace, or Descartes, properly if we understand them to be working with the same entities, problems, and questions. Both are doing physics. But Descartes considers physics to be continuous with theology, and considers problems about God's essence and will to be central to solving problems *for physics*, including problems about the necessity of the laws of nature. Laplace sees himself as having no need of a theology that is continuous with his physics, and so he constructs a physical system that does not appeal to the existence of God or even include any assertions about God. Laplace does not consider problems about God's essence and will to be problems in the domain of physics.

Can we give a reason, a *scientific* reason, why Descartes was *wrong* to include God as a "hypothesis" in his system of physics? We can certainly argue that Laplace's system is simpler, and thus argue on the basis of Ockham's razor.[32] Locally, we can judge Laplace's system from Descartes', and Descartes' from Laplace's. Neither allows for a knockdown argument, that the other must accept, why statements about the existence and attributes of God should or should not figure in physical proofs. On Kuhn's account, an explanation of why a scientist takes certain problems seriously, and others not, may be based on scientific reasons of one kind or another, or it may be based on local facts about the development of a research tradition, including explanations of scientists' behavior that depend on such local facts. Such local, specific explanations of scientific practice are of value for the history and philosophy of science. Above all, Kuhn's emphasis on textbook science and on the initiation of researchers into a specific, local tradition is salutary. It is a mistake simply to assume that scientists who have been trained differently will approach the phenomena in the same way, see the same problems as salient, and so on. We may not be able to understand events in the history of science properly if we do not pay attention to the training, pedagogy, and initiation of scientists.[33]

Researchers are made, and they are made with difficulty. That much is familiar to any working scientist or mathematician. The practical analogue of the paradigm inadequacy thesis is that, as a local and practical matter, no scientist can solve every problem using her current scientific training. Kuhn's practical account thus leads to the injunction that scientists should be aware of the existence of rival approaches, should learn about them, and should be aware of the limitations of their own approaches.

To be sure, Kuhn himself was quite pessimistic about the prospects for enlightened science along these lines, arguing that scientists in his experience were dogmatic and blinkered.[34] That does not, however, license *global* claims about the inability of researchers from rival or distinct paradigms to understand each other, or to work in a common world.

Kuhn's early work was criticized by philosophers who wished to see a more robust role for language and semantics in his view. Kuhn's work was even "Sneedified," as Damböck (2014) details, so that it fit into the formal, semantic tradition associated with Sneed and Stegmüller.[35] Along the way, Kuhn's statements about incommensurability came to be seen—even by Kuhn himself—as broad claims about lexical or taxonomic "speciation" between theories (Marcum 2015b), and as limitations on the ability of scientists even to express their results using rival conceptual frameworks.

Such developments are a shame, in my view. Kuhn's original work did not restrict "paradigm" to "theoretical framework," nor did he restrict the perspective of scientific practice to the content of propositions with a truth-value. And it is mainly because Kuhn's arguments in *Structure* are outside the semantic view, and focus instead on the practice of science, that they are interesting and fresh.

Rather than reading Kuhn through the lens of semantic theorists like Quine, Davidson, and Sneed, I would urge reading Kuhn's project in the lines of recent work on the "context of pedagogy" (Kaiser 2005, Richardson 2012, Woody 2004) and Hasok Chang's emphasis on historical understanding (Chang 2010). It is a long-standing project in the history and philosophy of science to understand, not only what scientists take themselves to be saying, but also "what the devil scientists thought they were up to" (Rudwick 1985, 12).

Understanding local practices in science, the importance of training and education, the salience of which problems researchers take to be compelling, and the shifts that take place as standards change with novel achievements and changes to the context, are all necessary to working out what scientists are doing and—just as importantly, from a historical perspective—what they think they are doing.

NOTES

1 I would like to thank Moti Mizrahi for his kind invitation to contribute to this volume, and for the provocative and compelling questions he has raised. James Collier first invited me to respond to Mizrahi's work in the *SERRC*, and this was a first occasion to think about these questions. My own work has benefited in large measure from the nuanced and well-argued contributions of Vasso Kindi and James Marcum to the exchange with Mizrahi. Barry Lam invited me to think through incommensurability, paradigms, and other central Kuhnian concepts, which has helped me to craft clearer descriptions of them. Alan Richardson made incisive comments on inchoate versions of those descriptions. Some of the research for this chapter was supported by, and done during a visit to, Martin Kusch's ERC project, "The Emergence of Relativism."

2 "Most of the sources cited in *Structure* are sources in the history of science (see Wray 2015). To be precise, 60% of the sources cited in *Structure* are in the history of science" (Wray 2016, 10).

3 By Masterson's count (1970), Kuhn uses "paradigm" in twenty-one ways in *Structure*. And later, Kuhn admits that his thinking changed. But this is the first clear *definition* Kuhn gives in *Structure*. Brorson and Andersen (2001) explain the early and continued influence of Fleck's "textbook science" on Kuhn, the 1950s onward.

4 Those who do put emphasis on practice include Rouse (1998, 2013), Andersen (2000), Brorson and Andersen (2001, including an excellent bibliography of related work), Hoyningen-Huene (1993, 2002), Richardson (2002), and others. As Rouse (2013, 59) says, "Kuhn's challenge to received philosophical views has been domesticated by reading him as offering an alternative conception of scientific knowledge. Kuhn is better understood as rejecting knowledge-centric accounts altogether, in favor of understanding the practice of research."

5 Richardson (2002) has referred to paradigms as giving rules of a game—paradigms guide scientific research. To be sure, Kuhn argued that no comprehensive rules that govern problem solving (and guarantee problems can be solved) could be given ahead of time. But paradigms can give rules for how to approach those problems.

6 Classic criticisms come in the essays in Lakatos and Musgrave (1970). Mayo (1996) is a later example.

7 Perhaps unacknowledged.

8 Timmins (2013) weighs the allegation that Kuhn plagiarized ideas from Polanyi's *Personal Knowledge* and the broader question of Polanyi's influence on *Structure*.

9 See Brorson and Andersen (2001, 110). Wray observes, "Kuhn refers to Fleck's *Genesis and Development of a Scientific Fact* in the preface to *Structure*, noting that it was instrumental in helping him see that his own project was tied to 'the sociology of the scientific community'" (Kuhn 2012/1962 xli; Wray 2016, 4).

10 Hoyningen-Huene (1993, 187), cited in Brorson and Andersen (2001, 110), who link this notion to their analysis of Fleck.

11 See also Hoyningen-Huene (2002), which also discusses Feyerabend in this connection.

12 This is a simplified form of an example Kuhn uses often.

13 Massimi (2015, 84–86) provides a detailed, more technical analysis of one of Kuhn's examples of incommensurability: Galileo's and Aristotle's treatment of falling bodies, before and after Galileo's discovery of the law of free fall.

14 A paraphrase of the commitments of causal descriptivists, including Kripke; for a recent discussion and references to further work, see Patton (2015a).

15 A paraphrase of a classic conception of scientific realism provided by Bas van Fraassen: "Science aims to give us, in its theories, a literally true story of what the world is like: and acceptance of a scientific theory involves the belief that it is true" (1980, 8, emphasis removed).

16 A phrase used by Lakatos, which he ascribes to Carnap. The picture described here is *not* the logical empiricist one, even though some details (like the public availability of phenomena) are found in that tradition.

17 Lakatos argued that paradigm shifts are rationally reconstructible (see, e.g., his chapter in Lakatos and Musgrave 1970). Patton (2012) argues for that claim in a qualified way based on Laudan's "context of pursuit," and provides a bibliography of work on this subject.

18 To be sure, Kuhn's critics are not all realists. I have to simplify a vast literature in this discussion.

19 The work of Friedman (2008, 2001), for instance, analyzes successive paradigms in physics in which one serves as a limiting case of another, which preserves "retrospective rationality." Paradigms are nested like Russian dolls, so that a later, more comprehensive paradigm can explain earlier ones.

20 Marjorie Grene to Kuhn, September 25, 1963; details in references.

21 Of course, it is possible that Kuhn would allow for a ground level of observation that does not depend on theory or on a given practical approach. But his account of *scientific* observation is not at that ground level.

22 Massimi makes clear that Kuhn does not argue for the ontological mind-dependence of the phenomena. Still, as Clark Glymour has recalled, and Norton conveys, "Clark and Hartry Field were having lunch in the cafeteria when Hartry remarked on Kuhn's curious view. When Thomson made his discovery with cathode rays (or however it was done), that's when Kuhn believes electrons popped into existence. Did Kuhn really think that?! At that moment, Kuhn just happened to walk by. Clark stopped him and asked. 'Yeah, of course,' Kuhn replied and he walked away" (Norton 2012). As Norton notes, this exchange does not answer the question of precisely what Kuhn meant. I would note, in particular, he may have thought the specific *semantic kind*, "electron," began to exist when it was experimentally demonstrable,

rather than that the *physical referent* of "electron" did not exist. Massimi provides an argument for a view resembling the former assertion.

23 Moti Mizrahi commented on a draft of this chapter that this point raises the question of how science could ever have started. For instance, when the "first astronomers" looked at the night sky, would they have had access to a paradigm? In *Structure*, Kuhn identifies "pre-paradigm science" as the initial phase of research (this idea is discussed throughout the work, including on pages 20, 48, 61, 162, and 178 of the edition cited). Pre-paradigm scientists are still working within an agreed-upon conceptual and practical framework, however. On the reading discussed here, it is that framework that allows them to have access to structured "phenomena" and not bare sense data, for instance.

24 It is entertaining to think in more detail about what a super-paradigm could be. For instance, robot scientists equipped with supercomputers might be the only earthly beings equipped to carry out scientific research under such a paradigm. But I will leave these speculations aside, reluctantly, for now.

25 There are more possible moves, of course, but these are prominent ones.

26 As Sankey (1993, 1997) makes clear, Kuhn's notions of incommensurability change over time. Marcum (2015b, 153) connects Kuhn's later emphasis on "changes in the lexical taxonomy of a scientific specialty" with Kuhn's "Darwinian" picture of science, so that "scientific progress is analogous to biological speciation, with incommensurability as the isolation mechanism."

27 Leplin (1986, 33) gives the example of Millikan's oil drop experiment: "if we describe what Millikan was doing without mentioning electrons, we seem to impute to him an unaccountable, indeed perverse interest in the amount of electric charge with which X-radiation will endow an oil droplet. What was the experiment for if not to determine the charge of the electron?"

28 As Sankey and Marcum have emphasized, in the works cited in the chapter.

29 I am grateful to Alan Richardson for emphasizing this passage and its significance. (He is, of course, not therefore responsible for my reading or use of the passage.)

30 Intriguingly, in the contemporary context, some realities of funding are pushing in the opposite direction, toward the questions that were considered by natural philosophers.

31 I would like to thank Moti Mizrahi for pressing clarification of this discussion, which has improved the account.

32 Anyone who tries to do this is not terribly familiar with Laplace's system.

33 Alan Richardson pointed out the salience of this question in the context of scientific practice.

34 Steve Fuller (2000) has reproached Kuhn on this score, arguing that Kuhn, in his alliance with Conant, bolstered Cold War science through his conservatism about science (see also Wray 2016). While a full response is beyond the scope of this chapter, I would note the following. Criticizing Kuhn for writing science as conservative and dogmatic is like asking Agatha Christie why she murdered all those people. Kuhn's assertions that science is conservative are descriptive, not normative: Kuhn's descriptions of dogmatic scientists are not flattering. For a more detailed and persuasive argument on this score, see Kindi (2003).

35 Kuhn himself spoke approvingly of Sneed and Stegmüller, but *only* in the context of reconstructing hierarchies of theoretical terms within successive theories, a context in which Kuhn also speaks approvingly of the "often elegant apparatus developed by the logical empiricists" (Kuhn 2000/1987, 14).

REFERENCES

Andersen, Hanne. 2000. "Learning by Ostension." *Science and Education* 9 (1): 91–106.

Brorson, Stig and Andersen, Hanne. 2001. "Stabilizing and Changing Phenomenal Worlds." *Journal for General Philosophy of Science/Zeitschrift für allgemeine Wissenschaftstheorie* 32 (1): 109–29.

Carnap, Rudolf, Morris, Charles, and Neurath, Otto (eds.). 1970. *Foundations of the Unity of Science. Towards an International Encyclopedia of Unified Science.* Vols. 1 and 2, Chicago: University of Chicago Press.

Chang, Hasok. 2010. "The Hidden History of Phlogiston." *HYLE* 16 (2): 47–79.

Damböck, Christian. 2014. "Caught in the Middle: Philosophy of Science Between the Historical Turn and Formal Philosophy as Illustrated by the Program of 'Kuhn Sneedified.'" *HOPOS* 4 (1): 62–82.

Friedman, Michael. 2001. *The Dynamics of Reason.* Stanford: C S L I Publications.

Friedman, Michael. 2008. "Ernst Cassirer and Thomas Kuhn." *The Philosophical Forum* 39 (2): 239–52.

Fuller, Steve. 2000. *Thomas Kuhn.* Chicago: University of Chicago Press.

Grene, Marjorie, Letter to Kuhn, September 25, 1963; Thomas S. Kuhn Papers, MC240, box 4; folder 9, Correspondence E-G; Massachusetts Institute of Technology, Institute Archives and Special Collections. Cited in Wray 2016.

Hoyningen-Huene, Paul. 1993. *Reconstructing Scientific Revolutions.* Chicago: University of Chicago Press.

Hoyningen-Huene, Paul. 2002. "Paul Feyerabend und Thomas Kuhn." *Journal for General Philosophy of Science/Zeitschrift für allgemeine Wissenschaftstheorie* 33 (1): 61–83.

Kaiser, David, ed. 2005. *Pedagogy and the Practice of Science: Historical and Contemporary Perspectives.* Cambridge, MA: MIT Press.

Kindi, Vasso. 2015. "The Role of Evidence in Judging Kuhn's Model." *Social Epistemology Review and Reply Collective* 4 (11): 25–33.

Kindi, Vasso. 2003. "Kuhn's Conservatism." *Social Epistemology* 17 (2 & 3): 209–14.

Kuhn, Thomas. 1977. *The Essential Tension.* Chicago: University of Chicago Press.

Kuhn, Thomas. 1959/1977. "The Essential Tension: Tradition and Innovation in Scientific Research." pp. 225–39 in Kuhn 1977.

Kuhn, Thomas. 2012/1962. *The Structure of Scientific Revolutions.* 50th anniversary edition. Chicago: University of Chicago Press.

Kuhn, Thomas. 2000/1987. "What Are Scientific Revolutions?" Reprinted in *The Road since Structure.* Chicago: University of Chicago Press.

Lakatos, Imre and Musgrave, Alan (eds.). 1970. *Criticism and the Growth of Knowledge*, Cambridge: Cambridge University Press.

Leplin, Jarrett. 1986. "Methodological Realism and Scientific Rationality." *Philosophy of Science* 53 (1): 31–51.

Marcum, James. 2015a. "What's the Support for Kuhn's Incommensurability Thesis?" *Social Epistemology Review and Reply Collective* 4 (9): 51–62.

Marcum, James. 2015b. *Thomas Kuhn's Revolutions: A Historical and an Evolutionary Philosophy of Science?* London: Bloomsbury.

Massimi, Michela. 2015. "'Working in a new world': Kuhn, Constructivism, and Mind-dependence." *Studies in History and Philosophy of Science* 50, 83–89.

Masterson, Margaret. 1970. "The Nature of a Paradigm." pp. 59–89 in Lakatos and Musgrave (eds.) 1970.

Mayo, Deborah. 1996. "Ducks, Rabbits, and Normal Science: Recasting the Kuhn's-Eye View of Popper's Demarcation of Science." *The British Journal for the Philosophy of Science* 47 (2): 271–90.

Mizrahi, Moti. 2013. "The Argument from Underconsideration and Relative Realism." *International Studies in the Philosophy of Science* 27 (4):393–407.

Mizrahi, Moti. 2015a. "Kuhn's Incommensurability Thesis: What's the Argument?" *Social Epistemology* 29 (4): 361–78.

Mizrahi, Moti. 2015b. "A Reply to Patton's 'Incommensurability and the Bonfire of the Meta-Theories.'" *Social Epistemology Review and Reply Collective* 4 (10): 51–53.

Mizrahi, Moti. 2015c. "A Reply to James Marcum's 'What's the Support for Kuhn's Incommensurability Thesis?'" *Social Epistemology Review and Reply Collective* 4 (11): 21–24.

Mizrahi, Moti. Forthcoming. "The History of Science as a Graveyard of Theories: A Philosophers' Myth?" *International Studies in the Philosophy of Science*.

Norton, John. 2012. Blog post, "Carnap and Kuhn: A Reappraisal." URL: <http://www.pitt.edu/~pittcntr/Being_here/last_donut/donut_2012-13/10-20-12 carnap kuhn.html>.

Patton, Lydia. 2012. "Experiment and Theory Building." *Synthese* 184 (3): 235–46.

Patton, Lydia. 2015a. "Methodological Realism and Modal Resourcefulness." *Synthese* 192 (11): 3443–62.

Patton, Lydia. 2015b. "Incommensurability and the Bonfire of the Meta Theories." *Social Epistemology Review and Reply Collective* 4 (7): 51–58.

Richardson, Alan. 2002. "Narrating the History of Reason Itself." *Perspectives on Science* 10 (3): 253–74.

Richardson, Alan. 2012. "Occasions for an Empirical History of Philosophy of Science." *HOPOS* 2 (1): 1–20.

Rouse, Joseph. 1998. "Kuhn and Scientific Practices." Division I Faculty Publications, Wesleyan University. Paper 17. http://wesscholar.wesleyan.edu/div1facpubs/17

Rouse, Joseph. 2013. "Recovering Thomas Kuhn." *Topoi* 32 (1): 59 64.

Rudwick, Martin. 1985. *The Great Devonian Controversy*. Chicago: The University of Chicago Press.

Sankey, Howard. 1993. "Kuhn's Changing Concept of Incommensurability." *The British Journal for the Philosophy of Science* 44 (4): 759–74.

Sankey, Howard. 1997. "Incommensurability: The Current State of Play." *Theoria* 12 (3): 425–45.

Timmins, Adam. 2013. "Why Was Kuhn's *Structure* More Successful Than Polanyi's *Personal Knowledge?*" *HOPOS* 3 (2): 306–17.

Toulmin, Stephen. 1967. "Conceptual Revolutions in Science." In *A Portrait of Twenty-five Years, Boston Colloquium for the Philosophy of Science 1960–1085.* Edited by R. S. Cohen and M. W. Wartofsky, 58–74. Dordrecht: D. Reidel Publishing Co.

Van Fraassen, Bas. 1980. *The Scientific Image.* Oxford: Oxford University Press.

Woody, Andrea. 2004. "More Telltale Signs." *Philosophy of Science* 71 (5): 780–93.

Wray, K. Brad. 2015. "Kuhn's Social Epistemology and the Sociology of Science." pp. 167–83 in *Kuhn's Structure of Scientific Revolutions: 50 Years On*, eds. William J. Devlin and Alisa Bokulich. Dordrecht: Springer.

Wray, K. Brad. 2016. "The Influence of James B. Conant on Kuhn's *Structure of Scientific Revolutions*." *HOPOS* 6 (1): 1–23.

Part III

REVISING THE KUHNIAN IMAGE
OF SCIENCE

Chapter 7

Redefining Revolutions[1]

Andrew Aberdein

1. A NICE KNOCK-DOWN ARGUMENT

Moti Mizrahi does an admirable job in pruning the thicket that has grown up around Thomas Kuhn's incommensurability thesis. He begins with a distinction between two versions of the thesis:

> **Taxonomic Incommensurability (TI)** Periods of scientific change (in particular, revolutionary change) that exhibit TI are scientific developments in which existing concepts are replaced with new concepts that are incompatible with the older concepts. The new concepts are incompatible with the old concepts in the following sense: two competing scientific theories are conceptually incompatible (or incommensurable) just in case they do not share the same "lexical taxonomy." A lexical taxonomy contains the structures and vocabulary that are used to state a theory.
>
> **Methodological Incommensurability (MI)** There are no objective criteria of theory evaluation. The familiar criteria of evaluation, such as simplicity and fruitfulness, are not a fixed set of rules. Rather, they vary with the currently dominant paradigm (Mizrahi 2015a, 362; references omitted).

Mizrahi's focus is exclusively on (TI), a focus that I will share. He proceeds to argue that, understood as (TI), the incommensurability thesis is poorly motivated. His critique differs from that of many other authors in focusing on the weakness of the arguments in support of (TI), rather than the strength of the arguments against it. He claims that the former arguments must be either deductive or inductive. In both cases, he presents a counterargument. Against deductive support, he argues as follows:

1. Reference change (discontinuity) is conclusive evidence for (TI) only if reference change (discontinuity) entails incompatibility of conceptual content.

133

2. Reference change (discontinuity) does not entail incompatibility of conceptual content.

Therefore:

3. It is not the case that reference change (discontinuity) is conclusive evidence for (TI) (Mizrahi 2015a, 367).

Against inductive support, he argues as follows:

1. There is a strong inductive argument for (TI) only if there are no rebutting defeaters against (TI).
2. There are rebutting defeaters against (TI).

Therefore:

3. It is not the case that there is a strong inductive argument for (TI) (Mizrahi 2015a, 371).

The merits of this critique have been addressed elsewhere (Kindi 2015; Marcum 2015; Patton 2015; Mizrahi 2015b, 2015c). In this chapter, I will take a somewhat different tack, by examining the implications for the philosophy of mathematical practice, specifically the debate whether there can be mathematical revolutions. Hence, I will not engage closely with all the details of Mizrahi's arguments. However, I do wish to draw attention to his use of "rebutting defeater." Before introducing an example from the history of medicine of a revolution in which conceptual continuity is displayed, he stresses that

the following episode is not supposed to be a counterexample against (TI). It is not meant to be a refutation of (TI). Rather, it shows that an inductive argument based on a few selected historical episodes of scientific change does not provide strong inductive support for (TI). Or, to put it another way, this episode—and others like it—counts as what Pollock calls a *rebutting defeater*, i.e. a *prima facie* reason to believe the negation of the original conclusion; in this case, the negation of (TI) (Mizrahi 2015a, 368; reference omitted).

That is, cases of revolutionary change without conceptual discontinuity are rebutting defeaters for (TI) since they are reasons to think that we can have one without the other. They are not undercutting defeaters, since the familiar cases of revolutionary change *with* conceptual discontinuity are still reasons to believe (TI), but if we have as many reasons to disbelieve it as to believe it, we should probably suspend our judgment.

2. THERE'S GLORY FOR YOU!

Alice's encounter with Humpty Dumpty is well-known to philosophers:

> "I don't know what you mean by 'glory,'" Alice said.
>
> Humpty Dumpty smiled contemptuously. "Of course you don't—till I tell you. I meant 'there's a nice knock-down argument for you!'"
>
> "But 'glory' doesn't mean 'a nice knock-down argument,'" Alice objected.
>
> "When *I* use a word," Humpty Dumpty said, in rather a scornful tone, "it means just what I choose it to mean—neither more nor less."
>
> "The question is," said Alice, "whether you *can* make words mean so many different things."
>
> "The question is," said Humpty Dumpty, "which is to be master—that's all" (Carroll 1897, 106 f.).

Several echoes of this passage may be heard in this chapter, most clearly in a Humpty-ish passage of my own:

> A *glorious* revolution occurs when the key components of a theory are pre-served, despite changes in their character and relative significance. (We will refer to such preservation, constitutive of a glorious revolution, as *glory*.) An *inglorious* revolution occurs when some key component(s) are lost, and perhaps other novel material is introduced by way of replacement . . . . A *paraglorious* revolution occurs when all the key components are preserved, as in a glorious revolution, but new key components are also added . . . . A *null* revolution[2] . . . when none of its key components change at all (Aberdein and Read 2009, 618 f).

Here is a slightly more formal characterization of this fourfold distinction. Let us specify that the key components of a theory T_n comprise a structure of some sort, K_n. Then succession between theories T_n and T_{n+1} would be a null revolution if their respective key components are unchanged, that is $K_n = K_{n+1}$, and a glorious revolution if the key components are in some sense isomor-phic, $K_n \cong K_{n+1}$, that is if there is a well-motivated bijection between them, which respects their roles in each theory. By contrast, in a paraglorious revo-lution there would be an analogous isomorphic embedding of the key terms of the old theory within the new, $K_n \hookrightarrow K_{n+1}$, such that the new theory contains terms with no clear counterpart in the old. And, in what we may call a strict inglorious revolution, there would be a similar isomorphic embedding of the key terms of the new theory within the old, $K_n \hookleftarrow K_{n+1}$, since there are compo-nents of the old theory that have been irretrievably lost in the new. In other words, inglorious revolutions would exhibit "Kuhn loss," a loss of (actual or potential) explanatory power (for further discussion, see Votsis 2011, 111 ff.). (The more general sense of inglorious revolution may be thought of as a strict inglorious revolution combined with a paraglorious revolution. That is, there would be some structure K'_n, not necessarily corresponding to any actually espoused theory, such that $K_n \leftrightarrow K'_n \hookrightarrow K_{n+1}$.)

In my earlier presentation of this distinction, I addressed a number of questions, the most important of which are what components are "key" and how are they "preserved"? The simplest answer to the first question would be to make *all* components of a theory key, or at least all components without which the theory could not be articulated. A more subtle account would permit distinctions between the theory proper and auxiliary theories, tentative extensions, and other inessential components, but we need not explore that account here.[3] Indeed, mathematical theories are less trouble than empirical theories in this respect: they characteristically have fewer components and their dependencies are much more clearly stated. The second question is much more of a challenge. Indeed, it is central to understanding what makes a revolution revolutionary: how much change is required for a revolution? Conversely, how much change can a theory undergo without revolution? In other words, just what do we mean by "glory"? An obvious starting point would be taxonomic commensurability; that is, the absence of (TI). Notice, incidentally, that lexical taxonomy is shared across neither inglorious nor paraglorious revolutions. I have defined inglorious and paraglorious revolutions such that the distinction is essentially a matter of historic sequence: whether such a transition counts as inglorious or paraglorious will depend on which theory came first. The definition of (TI), however, despite references to "new" and "old," does not seem to directly appeal to chronology.

Another question posed by this framework concerns the transitivity (or not) of these different characterizations of change. A succession of null revolutions must be a null revolution, because we have stipulated strict identity between key components. But a succession of glorious revolutions need not be a glorious revolution: a series of comparatively small changes might add up to a big change, as in a sorites sequence of small changes of color from red to blue. Likewise, the preservation aspect of paraglorious revolution may fail over a long enough sequence of such revolutions, making the sequence as a whole inglorious. Inglorious revolutions themselves are more straightforward: in principle, two consecutive inglorious revolutions might cancel each other out, so inglorious revolution must be intransitive.

A final concern leads us back to Humpty Dumpty: mere survival (or not) of vocabulary is not what is at issue.[4] Conceptual change might be disguised by a shift in the meaning of a shared vocabulary; conversely, a drastic change of vocabulary may give a misleading impression of change when nothing substantive has occurred. This issue is familiar from political examples: Augustus strategically reused much of the terminology of the old Roman Republic; Stalin was careful not to call himself a Tsar. In our context, this suggests that there are not four, but sixteen relationships between the key components of succeeding theories (using subscripts to indicate outward appearances):

$$K_n =_= K_{n+1} \qquad K_n \cong_= K_{n+1} \qquad K_n \hookrightarrow_= K_{n+1} \qquad K_n \hookleftarrow_= K_{n+1}$$
$$K_n =_\cong K_{n+1} \qquad K_n \cong_\cong K_{n+1} \qquad K_n \hookrightarrow_\cong K_{n+1} \qquad K_n \hookleftarrow_\cong K_{n+1}$$
$$K_n =_\hookrightarrow K_{n+1} \qquad K_n \cong_\hookrightarrow K_{n+1} \qquad K_n \hookrightarrow_\hookrightarrow K_{n+1} \qquad K_n \hookleftarrow_\hookrightarrow K_{n+1}$$
$$K_n =_\hookleftarrow K_{n+1} \qquad K_n \cong_\hookleftarrow K_{n+1} \qquad K_n \hookrightarrow_\hookleftarrow K_{n+1} \qquad K_n \hookleftarrow_\hookleftarrow K_{n+1}$$

The salutary point here is that we need to be careful in how we specify our terms, lest we misclassify (apparent) revolutions. In particular, the sharing of "lexical taxonomy" had better be more than just lexical, or the innocuous null revolution at the bottom left could count as a case of (TI), while the stealthy inglorious revolution at the top right does not.

A bold skeptical thesis would be that the rightmost two columns are in practice uninstantiated; that is, all revolutions are glorious, and all appearances to the contrary deceptive. As we will see in the next section, just such a view has been proposed by some historians of mathematics.

3. MATHEMATICAL REVOLUTIONS?

The discussion of mathematical revolutions essentially begins with Michael Crowe, who boldly asserts as a "law" that "Revolutions never occur in mathematics" (Crowe 1975, 19). Nonetheless, his own subsequent writings are increasingly nuanced: he has moved from denying that there are any revolutions in mathematics to suggesting that even inglorious revolutions may be possible (1988, 264 f.; 1992, 313). Joseph Dauben has published several articles arguing for the existence of mathematical revolutions (Dauben 1984, 1992, 1996). However, as we shall see, his position is much closer to Crowe's than might be expected. Indeed, as one commentator, writing more than twenty years ago, has remarked, the literature on mathematical revolutions represents an "authentic theoretical shambles" (Otero 1996, 193). There have been two collections of papers on the topic (Gillies 1992; Ausejo and Hormigón 1996). But the editor of the first notes that each of his authors "has a different theoretical perspective" (Gillies 1992, 8). And a contributor to the second pointedly observes that none of these authors make much use of a Kuhnian framework in their other writing on mathematical practice (Corry 1996, 170). In this section, I shall attempt to resolve some of this confusion.[5]

Many accounts of revolutions in mathematics distinguish two sorts of revolution, usually in terms of the presence or absence of some sort of conceptual continuity. Hence Crowe distinguishes a "transformational event," in which "an accepted theory is overthrown by another theory, which may be old or new," from a "formational event," in which "an area of science is not transformed, but is *formed*. The discovery that produces this effect is usually new, and by definition overthrows and replaces nothing" (Crowe 1992, 310, citing his own earlier work). Dauben likens mathematical revolutions to the Glorious Revolution of 1688, in the persistence of the "old order," albeit "under different terms, in radically altered or expanded contexts" (Dauben 1984, 52). This political analogy is echoed by Donald Gillies, who frames the distinction as between "Franco-British" and "Russian" revolutions: in the former, a "previously existing entity persists" through "a considerable loss of importance"; in the latter the "previously existing entity" is "overthrown and irrevocably discarded" (Gillies 1992, 5). Gillies' choice of terminology provokes the distracting historiographical question of why the French revolution should be more like the British than the Russian. After all, France and Russia both ended up as republics, whereas Britain did not. The answer seems to be that Gillies stops the clock at some point in the reign of Louis-Philippe. My own use of "glorious" revolutions, inspired by Dauben's usage but not intended to be given any specific historical reading, at least has the merit of sidestepping such musings. More importantly, we may notice that these distinctions do not necessarily coincide—Crowe's formational events seem closer to paraglorious than glorious revolutions, for example—and that, although all of them are binary, none of them seems to be exhaustive. So a more fine-grained distinction may be a source of clarity.

However the distinction is drawn, most of its framers agree that only glorious revolutions are possible in mathematics: "One important consequence, in fact, of the insistence on self-consistency within mathematics is that its advance is necessarily cumulative. New theories cannot displace the old" (Dauben 1984, 62); "in science both Russian and Franco-British revolutions occur. In mathematics, revolutions do occur but they are always of Franco-British type" (Gillies 1992, 6); "revolutions do occur in mathematics, but are confined entirely to the metamathematical component of the community's shared background" (Dunmore 1992, 223). In this regard, they concur with the opinion of many mathematicians that their discipline is cumulative. Crowe finds such sentiments expressed by Fourier in 1822, Hankel in 1869, and Truesdell in 1968 (Crowe 1975, 19). Indeed, celebrated mathematicians are still saying as much: "central contributions have been lasting, one does not supersede another, it enlarges it" (Langlands 2013, 25). As Crowe poetically summarizes the conventional view, "Scattered over the landscape of the past of mathematics are numerous citadels, once proudly erected, but which, although never attacked, are now left unoccupied by active mathematicians" (Crowe 1988, 263). Nonetheless, there are exceptions to this trend. Michael Harris notes Kronecker in 1891 observing that "in this respect mathematics is no different from the natural sciences: new phenomena overturn the old hypotheses and put others in their place" and Siegel in 1964 characterizing work revisionary of his own as "a pig broken into a beautiful garden and rooting up all flowers and trees" (Harris 2015, 4).

Bruce Pourciau complains that the "Crowe–Dauben debate" is actually a "Crowe–Dauben consensus," for example: Kuhnian revolutions are inherently impossible in mathematics (Pourciau 2000, 301). For Pourciau, a revolution is Kuhnian or "*noncumulative* whenever some true statements of the old conception have no translations (faithful to the original meaning) which are true statements in the new conception" (Pourciau 2000, 301). Pourciau argues that Brouwerian intuitionism is a (failed) Kuhnian revolution. Certainly, if adopted, this would have required wholesale revision of results treated as certain by prior mathematicians, thereby meeting the strictest definition of Kuhnian revolution. Its usefulness as an example might be somewhat compromised by the fact that it never actually happened, but it was seriously proposed and still commands some support. However, Pourciau may be overestimating the difficulty in supplying examples of Kuhnian revolutions in mathematics in two ways: one of scale, one of chronology. Firstly, although the standard example of a Kuhnian revolution in natural science is the Copernican revolution, an epochal upending of an all-encompassing worldview, it is a mistake to suppose that all Kuhnian revolutions need be so drastic. Stephen Toulmin once complained that Kuhn had surreptitiously revised his position to admit "small-scale 'micro-revolutions'" (Toulmin 1970, 47). Kuhn strenuously rejected this imputation: "My concern . . . has been throughout what Toulmin now takes it to have become: a little studied type of conceptual change, which occurs frequently in science and is fundamental to its advance" (Kuhn 1970, 249 f.). He subsequently characterized a paradigm as "what the members of a scientific community and they alone share" where such communities may comprise "perhaps 100 members, sometimes significantly fewer" (Kuhn 1974, 460; 462). Happily enough, this coincides with an influential estimate of the size of mathematical research communities: "a few dozen (at most a few hundred)" (Davis and Hersh 1980, 35). Secondly, as observed earlier in the chapter, paraglorious and inglorious revolutions are essentially symmetrical; they differ only in the chronological sequence of the contrasting theories. Paraglorious revolutions are cumulative, but they exhibit a conceptual discontinuity formally identical to that exhibited by inglorious revolutions. Chronological sequence alone does not seem to be a principled basis on which to discount paraglorious revolutions as Kuhnian.[6]

Three broad strategies for the identification of Kuhnian (or other than glorious) revolutions in mathematics arise from this discussion. Firstly, we may look directly for inglorious revolutions: conceptual shifts within mathematics in which key components have been lost. Secondly, we may look for paraglorious revolutions: conceptual shifts within mathematics in which key components have been gained. Thirdly, we may look for sorites-like sequences of glorious (or paraglorious) revolutions that exhibit nontransitivity of glory; that is, which are collectively inglorious. The search is complicated by several factors. In particular, it is not easy to determine whether a given episode is revolutionary; nor is it easy to determine what type of revolution a given revolutionary episode exemplifies. Hence, some of the same examples might be claimed as successes for more than one of these search strategies. An analysis of even a single case study thorough enough to settle all of these issues would be beyond the scope of this chapter. However, in the following sections, I will discuss several putative mathematical revolutions in what I hope to be at least enough detail to indicate the prospects for these strategies.

4. $\mathbb{Q} \rightarrow \mathbb{R} \rightarrow \mathbb{C}$

The most obvious example of incommensurability in mathematics must be incommensurability itself! The concept is, of course, originally a mathematical one, credited to the very earliest Greek geometers, the Pythagoreans. Thomas Heath, in his edition of Euclid, quotes a scholium on the first proposition of Book X, attributed to Proclus: "They called all magnitudes measurable by the same measure commensurable, but those which are not subject to the same measure incommensurable" (Heath 2006 [1908], 684). Specifically, the Pythagoreans discovered that $\sqrt{2}$ was incommensurable with the natural numbers, that is, it cannot be expressed as a ratio of natural numbers or, as we would say, as a rational number. The discovery was credited to one Hippasus of Metapontum, who is reputed to have drowned in a shipwreck. As the historian of mathematics Kurt von Fritz observes,

> The discovery of incommensurability must have made an enormous impression in Pythagorean circles because it destroyed with one stroke the belief that everything could be expressed in integers, on which the whole Pythagorean philosophy up to then had been based. This impression is clearly reflected in those legends which say that Hippasus was punished by the gods for having made public his terrible discovery (Fritz 1945, 260).

Even in ancient times, the allegorical aspects of this story were already apparent, "hinting that everything irrational and formless is properly concealed, and, if any soul should rashly invade this region of life and lay it open, it would be carried away into the sea of becoming and be overwhelmed by its unresting currents," as Proclus puts it (Heath 2006 [1908], 684).

For our purposes, the crucial point in this narrative is that the change initiated by Hippasus was revisionary of earlier mathematics: it "required an entirely new concept of ratio and proportion and a new criterion to determine whether two pairs of magnitudes which are incommensurable with one another have the same [ratio]" (Fritz 1945, 262). The completion of this task by later mathematicians, notably Theaetetus and Eudoxus, is one of the great achievements of Greek mathematics, and plausibly a major driver of its early development of the concept of rigorous proof. For Dauben, it is one of the best examples of a mathematical revolution. He stresses that the "transformation of the concept of number . . . entailed more than just extending the old concept of number by adding on the irrationals—the entire concept of number was inherently changed, transmuted as it were, from a world-view in which integers alone were numbers, to a view of number that was eventually related to the completeness of the entire system of real numbers" (Dauben 1984, 57).

In my terminology, this is clearly not a glorious revolution, since a literally incommensurable concept has been added. So it is at least a paraglorious revolution. Might we go further and identify it as inglorious? On the one hand, something has certainly been lost: the "world-view in which integers alone were numbers," for a start. On the other hand, world-views are not part of the subject matter of mathematics. Hence, Caroline Dunmore identifies this shift as "the first great meta-level revolution in the development of mathematics" (Dunmore 1992, 215). On Dunmore's account, object level revolutions in mathematics are always glorious, but they are always accompanied by inglorious revolutions in the meta-level, that is in the philosophical or methodological presuppositions (Dunmore 1992, 225). The rational numbers are still an object of mathematical enquiry and the Pythagorean results about their comparison still hold. Nonetheless, it is highly misleading to conceive of the real numbers as a conservative extension of the rationals. The real numbers are constructed on a quite different basis, but in such a way that a subset isomorphic to the rationals may be identified.

Strictly speaking, the taxonomic incommensurability between mathematics defined over $\mathbb{Q}$ and mathematics defined over $\mathbb{R}$ runs both ways. Clearly, real mathematics cannot be done with rationals alone, but techniques that work over $\mathbb{Q}$ fail over $\mathbb{R}$. So a revolution from the mathematics of $\mathbb{Q}$ to the mathematics of $\mathbb{R}$ would be paraglorious and inglorious. However, the actual revolution could also be described as a shift from the mathematics of $\mathbb{Q}$ to (eventually) the mathematics of $\mathbb{Q}$ and $\mathbb{R}$, understood as separate projects. That shift would be strictly paraglorious—assuming that the mathematics of $\mathbb{Q}$ has been preserved, and not just reconstructed. The same issue arises with supersets of the reals, whether well established, such as the complex numbers, or more contentious, such as the hyperreals, which include infinitesimals (Bair et al. 2013). The underlying issue of cross-sortal identity is a known problem for a wide range of philosophies of mathematics (Cook and Ebert 2005, 124). Textbook presentations of the foundations of mathematics are obliged to address cross-sortal identity, which they do in a variety of ways, often at odds with mathematical practice (for a careful discussion, see Ganesalingam 2013, 180 ff.). It is also important to note that retrofitting a new foundation onto existing mathematics is not confined to number systems. Indeed, it has been a major feature of mathematical research since the nineteenth century—and it is precisely what Siegel was complaining about as "rooting up all flowers and trees" (Lang 1994, 22). It is a deep question whether such moves can be understood as merely adding "a new storey to the old structure" (Crowe 1975, 19, quoting Hankel). To pursue the architectural metaphor, they might be better characterized as "façading," whereby the front elevation of an otherwise demolished building is incorporated into its successor. I cannot settle that question here, but we may observe that these shifts are at least paraglorious and perhaps inglorious.

5. IRONY FOR MATHEMATICIANS

One way of approaching the issue of inglorious revolution in mathematics is through a related question: When do mathematicians say things that are not so? One prospect might be the assumptions of reductio proofs. In a recent essay, the mathematician Timothy Gowers briefly considers, but ultimately rejects, the intriguing idea "that proofs by contradiction are the mathematician's version of irony" (Gowers 2012, 224). He objects that "when we give a proof by contradiction, we make it very clear that we are discussing a counterfactual, so our words *are* intended to be taken at face value" (op. cit.). Perhaps more tellingly, we might frame this objection as saying that a proposition assumed for the purposes of proof by contradiction is presented as the antecedent of a conditional: "If P were the case, then . . . a contradiction would follow. So, not P." Nonetheless, at least for the duration of the ellipsis, the mathematician proceeds as though P were being seriously entertained. An inattentive reader who began reading a proof part way through would not necessarily be able to tell which proposition the mathematician intended to show to be false—or even that any of them were presented with this intent.

Another possibility might be unproven conjectures upon which mathematicians sometimes rely, when exploring their consequences. The mathematician Barry Mazur talks of "architectural conjectures" that "play the role of 'joists' and 'supporting beams' for some larger mathematical structure yet to be made" (Mazur 1997, 199). The formulation of such conjectures

> is often a way of "formally" packaging, or at least acknowledging, an otherwise shapeless body of mathematical experience that points to their truth . . . . These

conjectures sometimes round out a field by being clear, general (but not yet proved) statements enabling one to understand where a certain amount of ongoing, perhaps fragmentary, specialized work is headed; they provide a focus (op. cit.).

Architectural conjectures seldom arise alone; they often comprise elaborate networks of interlinked conjectures that present the outline of what is hoped to be many years of fruitful work. One of the best known such networks of conjectures in contemporary mathematics is the Langlands program, "an extensive web of conjectures by which number theory, algebra, and analysis are interrelated in a precise manner, eliminating the official divisions between the subdisciplines" (Zalamea 2012, 180). This has been enormously influential, guiding the work of scores of mathematicians who have confirmed some—but by no means all—of its key conjectures.

As with reductio hypotheses, conjectures are strictly to be understood as the antecedents of conditionals. Mathematicians should not be seen as *asserting* them until they have actually been proven. Nonetheless, as with reductio hypotheses, they are presented in apparent earnest, and their implications investigated with all due rigor: they "are *expected* to turn out to be true, as, of course, are all conjectures" (Mazur 1997, 199). So, naively, it may seem as though neither sort of hypothesis is much use for present purposes, since mathematicians seem to have an uncanny knack of only assuming for proof by contradiction things that are false and only assuming as conjectures things that are true (even if not yet proven). This would be a profound misperception: attempted reductios sometimes founder on the truth of the hypothesis (famously so in the accidental discovery of non-Euclidean geometry) and sometimes substantial effort is devoted to exploring the consequences of false conjectures. In the next section, we will encounter an example of the latter.

6. THE WORLD WITHOUT END HYPOTHESIS

In 2016, Michael Hill, Michael Hopkins, and Douglas Ravenel published an article in one of the most prestigious journals in mathematics, with the following abstract:

We show that the Kervaire invariant one elements $\theta_j \in \pi_{2^{j+1}-2}S^0$ exist only for $j \leq 6$. By Browder's Theorem, this means that smooth framed manifolds of Kervaire invariant one exist only in dimensions 2, 6, 14, 30, 62, and possibly 126. Except for dimension 126 this resolves a longstanding problem in algebraic topology (Hill et al. 2016, 1).

This was the final, fully vetted version of a result that they had announced seven years earlier. As they summarize their result in a preliminary expository article, they showed that "certain long sought hypothetical maps [the θ_j for $j{\geq}7$] between high dimensional spheres do not exist" (Hill et al. 2010, 32). This outcome was a surprising one, not just because of the technical depth of the work required but because many experts in the area had long expected the opposite result: "The problem solved by our theorem is nearly 50 years old. There were several unsuccessful attempts to solve it in the 1970s. They were all aimed at proving the opposite of what we have proved" (Hill et al. 2010, 32). The hypothesis that the sought after maps all exist came to be known as the World Without End Hypothesis; the contradictory hypothesis that the θ_j only exist for small j was known as the Doomsday Hypothesis. Hill, Hopkins, and Ravenel proved the Doomsday Hypothesis, and thereby disproved the World Without End Hypothesis.

While not remotely on the scale of the Langlands Program, the World Without End Hypothesis was not just a single assertion but the basis for a whole system of "architectural conjectures": the new proof demolished "what Ravenel calls an entire 'cosmology' of conjectures" (Klarreich 2011, 374). The triumph of the Doomsday Hypothesis undercut a growing sense of understanding provided by the World Without End Hypothesis. As the reviewer of Hill, Hopkins, and Ravenel's paper in *Mathematical Reviews* comments, the World Without End Hypothesis "was so compelling that many believed the θ_j must exist; now that we know they don't, the behavior of the EHP sequence is much more mysterious. In particular, Mahowald's η_j-elements . . . now appear entirely anomalous" (Goerss 2016). The author of a book surveying some of the techniques developed in pursuit of the World Without End Hypothesis published shortly before Hill, Hopkins, and Ravenel's announcement concluded his preface as follows: "In the light of the above conjecture [the Doomsday Hypothesis] and the failure over fifty years to construct framed manifolds of Arf-Kervaire invariant one this might turn out to be a book about things which do not exist" (Snaith 2009, ix).

The collapse of the World Without End Hypothesis seems to be an inglorious revolution. It is clearly a case of referential discontinuity, since a whole class of objects that were key to the old theory has been shown not to exist. It might be objected that there is no taxonomic incommensurability, since the conjectures are still perfectly intelligible, despite now being known to be false. Indeed, we have seen that Mizrahi argues that referential discontinuity does not entail conceptual incompatibility, and is thereby insufficient for taxonomic incommensurability (see chapter 1). So what else is required? A natural candidate would be Kuhn loss: reduction in actual or potential explanatory power. Kuhn loss does not seem to feature in the examples Mizrahi adduces of referential discontinuity without conceptual incompatibility (Mizrahi 2015a, 365 ff.), but it is exhibited here: the understanding that had seemed to follow from the World Without End Hypothesis has been lost. Specifically, comments such as Paul Goerss' reveal the "mysterious" and "anomalous" nature of some of the surviving results, now that the architectural conjectures that mathematicians had relied on to understand them have been falsified. This is what we should expect from a reduction in explanatory power.

If the collapse of the World Without End Hypothesis is a revolution, it is certainly a small-scale one. The constituency of homotopy theorists for whom this was an active research area would be at the low end of Kuhn's "perhaps 100 members, sometimes significantly fewer" comprising a scientific community (Kuhn 1974, 462). However, this makes this revolution much more likely to be typical of mathematical revolutions. Large-scale revolutions are necessarily extremely rare; so much so that their instances may well be *sui generis*. Conversely, small-scale revolutions are much better placed to support generalizations: there are plenty of other failed architectural conjectures that could be explored in similar detail (see, e.g., some of the "cautionary tales" in Jaffe and Quinn 1993, 7 f.).

7. INTER-UNIVERSAL TEICHMÜLLER THEORY

One of the more widely discussed open problems in modern number theory is the *abc* conjecture. Like many such problems, it can be stated quite simply but defies simple solution. Here is what it says:

> **The *abc* conjecture.** For every $\varepsilon > 0$, there are only finitely many triples (a,b,c) of coprime positive integers where $a+b=c$, such that $c > d^{1+\varepsilon}$, where d denotes the radical of *abc* (the product of its distinct prime factors).

For example, try $a=15$ and $b=28$. These are coprime, but $c=43$ and $d=2\times3\times5\times7\times43=9030 \gg 43$. So $(15,28,43)$ is not one of the specified triples (for any ε). On the other hand, let $a=1$ and $b=63$. Then we have $c=64$ and $d=2\times3\times7=42<64$. So $(1,63,64)$ is such a triple (at least for values of $\varepsilon < .11269$).

In a series of preprints appearing on his website in 2012, the respected mathematician Shinichi Mochizuki claimed to have a proof of the *abc* conjecture. However, Mochizuki's claimed proof introduced so many new techniques and concepts that other leading mathematicians in the field described it as like "reading a paper from the future, or from outer space" and as "very, very weird" (cited in Chen 2013). The scale of the proof (more than 500 pages) and its sheer incomprehensibility, even by the standards of cutting-edge research mathematics, have so far stalled all attempts at the normal processes of confirmation and acceptance that transform a proof claim into an established proof. Although a handful of other mathematicians now profess to understand Mochizuki's work, they have had little success sharing that understanding more widely. One anonymous mathematician, quoted in *Nature*, summed up the problem:

> "Everybody who I'm aware of who's come close to this stuff is quite reasonable, but afterwards they become incapable of communicating it". . . . The situation, he says, reminds him of the *Monty Python* skit about a writer who jots down the world's funniest joke. Anyone who reads it dies from laughing and can never relate it to anyone else (Castelvecchi 2015, 181).

Mochizuki calls his work inter-universal Teichmüller theory (IUTeich). He has reflected on the verification process of IUTeich in a pair of papers that comprise a valuable resource for philosophers of mathematical practice (Mochizuki 2013, 2014).[7] Mochizuki warns that "any attempt to study IUTeich under the expectation that the *essential thrust* of IUTeich will proceed via a similar pattern of argument to existing mathematical theories is likely to end in failure" (Mochizuki 2013, 5; all emphases Mochizuki's). Even Teichmüller theory itself is only an indirect inspiration. Nonetheless, IUTeich does echo Teichmüller theory in at least one respect—the papers in which Oswald Teichmüller laid out his theory were not immediately accepted by the mathematical community either: "It was after several years of hard work by several mathematicians that all the arguments in these papers were considered as being sound" (Ji and Papadopoulos 2013, 128). So a lengthy gap before final community acceptance is not unusual in itself. (Note also the seven years between initial announcement and final publication of Hill, Hopkins, and Ravenel's work.) What is unusual in Mochizuki's case is that the mathematical community appears to be completely stumped.

The trouble arises from both the scale and the nature of the task required of mathematicians who wish to come to terms with Mochizuki's work. He suggests, perhaps optimistically, that "it is quite *possible* to achieve a *reasonably rigorous understanding* of the theory within a period of *a little less than half a year*" (Mochizuki 2013, 4). But this is still a substantial investment of time. Mochizuki also notes that his work is essentially independent of the Langlands Program (discussed in section 5) (Mochizuki 2014, 10). Since this has guided so much recent work in number theory, many of the individuals most interested in the *abc* conjecture have a background that does not particularly suit them for tackling IUTeich, and should not necessarily expect to acquire techniques that would further their own projects from the six months or more of concentrated intellectual effort required. Indeed, Mochizuki stresses the incompatibility of the ideas behind IUTeich and the ideas most number theorists are familiar with: "the most essential stumbling block lies not so much in the need for the *acquisition of new knowledge*, but rather in the need for researchers . . . to *deactivate the thought patterns* that they have installed in their brains and taken for granted for so many years and then to start afresh" (Mochizuki 2014, 11 f.). He complains that

> when a researcher with a solid track record in mathematical research decides to read a mathematical paper, . . . such a researcher will attempt to digest the content of the paper in as efficient a way as is possible, by *scanning* the paper for important terms and theorems so that the researcher may apply his/her vast store of expertise and deep understanding of the subject to determine just *which* of those topics of the subject that, from point of view of the researcher, have already been "digested" and "well understood" *play a key role in the paper* . . . . Of course, in the case of IUTeich, a researcher who already possesses a deep understanding, as well as a solid track record in mathematical research, concerning such topics as absolute anabelian geometry, the rigidity properties of the étale theta function, and Hodge-Arakelov theory, may indeed find such "occasional nibbling" to be more than sufficient to attain a quite genuine understanding of IUTeich. In fact, however, for better or worse, *no such researcher exists* (other than myself) at the present time (Mochizuki 2014, 8 f).

This is an eloquent description of conceptual incompatibility. Mochizuki is not just saying that no one is capable of understanding IUTeich; on the contrary, he is confident that the material is well within the grasp of competent research mathematicians. But he stresses that, if they are to understand IUTeich, they must set aside their existing conceptual frameworks and build a new one from scratch. The contrast that Mochizuki draws echoes that Kuhn draws between translation and language acquisition; incommensurability being a bar to the former, but not the latter (Kuhn 2000 [1983], 53).

If the revolution that Mochizuki has so far failed to ignite succeeds, it would appear to be strictly paraglorious in nature. He does not wish to overturn anything; rather he wishes to comprehensively supplement the existing apparatus of number theory. If IUTeich is correct, it will represent a substantial leap forward in mathematics. Mochizuki's problem is that he is trying to do it all in one go. Conversely, if an irreparable flaw is found in Mochizuki's reasoning and IUTeich collapses, then its fall would be an inglorious revolution. In either case, IUTeich would exhibit similar conceptual incompatibility with mainstream number theory.

8. CLASSICAL→MODERN→CONTEMPORARY

As a final example, I wish to shift focus from some ultimately quite small-scale revolutions (albeit ones that have attracted a fair bit of publicity) to the

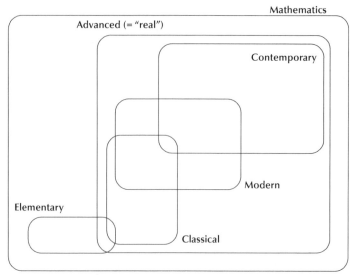

Mathematics

Advanced (= "real")

Contemporary

Modern

Elementary

Classical

Figure 7.1. Correlations between the areas of mathematics: elementary, advanced, classical, modern, contemporary. (© Fernando Zalamea, *Synthetic Philosophy of Contemporary Mathematics*, 2012. Used with permission.)

discipline of mathematics as a whole. The mathematician Fernando Zalamea offers the following useful periodization of research in his discipline:

> **Classical mathematics (midseventeenth to midnineteenth centuries):** sophisticated use of the infinite (Pascal, Leibniz, Euler, Gauss);
> **Modern mathematics (midnineteenth to midtwentieth centuries):** sophisticated use of structural and qualitative properties (Galois, Riemann, Hilbert);
> **Contemporary mathematics (midtwentieth century to present):** sophisticated use of the properties of transference, reflection and gluing (Grothendieck, Serre, Shelah) (Zalamea 2012, 27).

The scale of mathematical research grows with each generation, such that each period attacks a broader front and produces more results than its predecessors. Zalamea graphically represents this process in the diagram reproduced as figure 7.1. The moral for the philosopher of mathematical practice is a striking one: classical and modern mathematics may be familiar enough from school or undergraduate study, but contemporary mathematics almost certainly is not. Even philosophers who take pains to reflect on more than just elementary and foundational work may well be quite out of touch with the conceptual underpinnings of mathematical research conducted in their own lifetimes. Hence, as Zalamea complains, the large-scale conceptual shift from modern to contemporary mathematics has gone largely unremarked by philosophers. I would contend that this shift may be understood as revolutionary. Certainly, the "properties of transference, reflection and gluing" would be impossible to articulate with only the conceptual resources available to Galois, Riemann, or Hilbert (let alone Pascal, Leibniz, Euler, or Gauss). Insofar as these are key components of contemporary mathematics, their acquisition is at least paraglorious. Furthermore, contemporary mathematics has undergone a sorites-like sequence of paraglorious revolutions of such daunting scope that the key components preserved throughout the sequence have a drastically diminished role in the new era. So much so indeed, that the whole transition might best be characterized as inglorious.

9. CONCLUSION

To take stock, we have seen four case studies exemplifying different classes of putative revolution in mathematics: the shift from rational to real numbers (and other cases of foundational retrofitting); shifts occasioned by the collapse of an architectural conjecture, such as the World Without End Hypothesis; shifts resulting from a rapid advance, such as IUTeich; and the collective large-scale shift that has transformed recent mathematics. The first of these is at least paraglorious and perhaps also inglorious. The second seems to be strictly inglorious, whereas the third is strictly paraglorious (if successful; if unsuccessful, it would be another failed architectural conjecture). Both of these examples are comparatively small scale and might be seen as exemplary of similar shifts in other areas of mathematics. Lastly, the shift from modern to contemporary mathematics has involved numerous conceptual innovations, each of which might be seen as initiating a paraglorious revolution, and, when taken collectively, might represent a sorites-like inglorious revolution.

So where do Mizrahi and I agree and disagree? He disputes whether there are revolutions exhibiting (TI) in science; I have argued that such revolutions can be found in mathematics. I take it that we both agree with Crowe that it is a misconception that "the methodology of mathematics is radically different from that of science" (Crowe 1988, 271). So we should both like for the story we tell about revolutions to hold for both science and mathematics. Of course, you don't always get what you want—conventional wisdom might suggest that we are both wrong across the board: science and mathematics are methodologically discontinuous in part because science exhibits (inglorious) revolutions but mathematics does not. Conversely, someone might defend the contrarian stance that Mizrahi and I are both right about revolutions but wrong about the methodological continuity of science and mathematics, since (inglorious) revolutions are confined to mathematics. (This is not as absurd as it may appear—some of the strategies for minimizing the revolutionary aspects of conceptual shifts in science, such as finding common referents between theories, may not work in a field where all the referents are abstract objects.) However, I believe a more satisfactory resolution is possible.

To see how this might be accomplished, it will help to recast Mizrahi's arguments in my terminology. (TI) may be understood as saying that no revolutions are glorious. So a rebutting defeater against (TI) would be a glorious revolution. Since there are glorious revolutions, Mizrahi concludes that (TI) lacks strong inductive support. However, if we restrict (TI) to inglorious and paraglorious revolutions, then glorious revolutions no longer count as rebutting defeaters. Mizrahi does consider, and reject, a related proposal from the "friends of (TI)": retreating to the claim that "some episodes of scientific change exhibit TI, whereas others do not" (Mizrahi 2015a, 372). He rightly objects that such a claim would have no explanatory or predictive value. However, my proposal is more robust: rather than just exclude the anomalous cases, I have sketched an independent characterization of subtypes of revolution for which (TI) still holds.[8] Hence, Mizrahi is right that (TI) is false as a claim about revolutions in general. But (TI) is true of two important subtypes: paraglorious and inglorious revolutions.

NOTES

1 I am grateful to Moti Mizrahi for the invitation to contribute to this volume and for his insightful comments. I presented an earlier draft at the University of Nevada, Las Vegas: my thanks for a valuable discussion to the audience, and especially Ian Dove, Maurice Finocchiaro, and James Woodbridge.

2 Less happily, I also referred to theories in null revolution as being in stasis, but 'stasis' is a false friend: although it has come to mean an absence of change, for Aristotle, it meant something much like revolution, and is often translated as such (Howell 1985, 18).

3 The sort of distinction I have in mind is that Stathis Psillos draws between idle and essentially contributing constituents (Psillos 1996, S311) or Philip Kitcher between presuppositional and working posits (Kitcher 1993, 149).

4 Mizrahi also addresses this concern (Mizrahi 2015a, 374).

5 For a more extensive account of the debate over revolutions in mathematics, see (François and Van Bendegem 2010, 107 ff.).

6 I make no claim as to Kuhn's own view on this issue, although I note that he does refer to historians experiencing revolutions by "moving through time in a direction opposite to scientists" (Kuhn 2000 [1983], 57), which at least suggests an openness to temporal symmetry.

7 For a very different application of these papers to the philosophy of mathematical practice, see (Tanswell 2016, 187 ff.).

8 The contrast is akin to that Imre Lakatos draws between monster-barring and exception-barring (Lakatos 1976, 29).

REFERENCES

Aberdein, Andrew, and Stephen Read. 2009. "The Philosophy of Alternative Logics." In *The Development of Modern Logic*, edited by Leila Haaparanta, 613–723. Oxford: Oxford University Press.

Ausejo, Elena, and Mariano Hormigón, eds. 1996. *Paradigms and Mathematics.* Madrid: Siglo XXI Editores.

Bair, Jacques, Piotr Błaszczyk, Robert Ely, Valérie Henry, Vladimir Kanovei, Karin U. Katz, Mikhail G. Katz, Semen S. Kutateladze, Thomas McGaffey, David M. Schaps, David Sherry, and Steven Shnider. 2013. "Is Mathematical History Written by the Victors?" *Notices of the AMS* 60: 886–904.

Carroll, Lewis. 1897. *Through the Looking-Glass and What Alice Found There.* London: Ward Lock.

Castelvecchi, Davide. 2015. "The Biggest Mystery in Mathematics: Shinichi Mochizuki and the Impenetrable Proof." *Nature* 526(7572): 178–81.

Chen, Caroline. 2013. "The Paradox of the Proof." *Project Wordsworth.* Online at projectwordsworth.com/the-paradox-of-the-proof/.

Cook, Roy T., and Philip Ebert. 2005. "Abstraction and Identity." *Dialectica* 59: 121–39.

Corry, Leo. 1996. "Paradigms and Paradigmatic Change in the History of Mathematics." In *Paradigms and Mathematics*, edited by Elena Ausejo and Mariano Hormigón, 169–91. Madrid: Siglo XXI Editores.

Crowe, Michael J. 1975. "Ten 'Laws' Concerning Patterns of Change in the History of Mathematics." *Historia Mathematica* 2: 161–66. Reprinted in Gillies 1992, 15–20.

Crowe, Michael J. 1988. "Ten Misconceptions Concerning Mathematics and Its History." *Minnesota Studies in the Philosophy of Science* 11: 260–77.

Crowe, Michael J. 1992. "Afterword (1992): A Revolution in the Historiography of Mathematics?" In *Revolutions in Mathematics*, edited by Donald Gillies, 306–16. Oxford: Clarendon.

Dauben, Joseph Warren. 1984. "Conceptual Revolutions and the History of Mathematics: Two Studies in the Growth of Knowledge." In *Transformation and Tradition in the Sciences: Essays in Honor of I. Bernard Cohen*, edited by Elliott Mendelson, 81–103. Cambridge: Cambridge University Press. Reprinted in Gillies 1992, 49–71.

Dauben, Joseph Warren. 1992. "Appendix (1992): Revolutions Revisited." In *Revolutions in Mathematics*, edited by Donald Gillies, 72–82. Oxford: Clarendon.

Dauben, Joseph Warren. 1996. "Paradigms and Proofs: How Revolutions Transform Mathematics." In *Paradigms and Mathematics*, edited by Elena Ausejo and Mariano Hormigón, 117–48. Madrid: Siglo XXI Editores.

Davis, Philip J., and Reuben Hersh. 1980. *The Mathematical Experience.* Basel: Birkhäuser.

Dunmore, Caroline. 1992. "Meta-Level Revolutions in Mathematics." In *Revolutions in Mathematics*, edited by Donald Gillies, 209–25. Oxford: Clarendon.

François, Karen, and Jean Paul Van Bendegem. 2010. "Revolutions in Mathematics. More Than Thirty Years after Crowe's 'Ten Laws'. A New Interpretation." In

PhiMSAMP. Philosophy of Mathematics: Sociological Aspects and Mathematical Practice, edited by Benedikt Löwe and Thomas Müller, 107–20. London: College Publications.

Fritz, Kurt von. 1945. "The Discovery of Incommensurability by Hippasus of Metapontum." *Annals of Mathematics* 46: 242–64.

Ganesalingam, Mohan. 2013. *The Language of Mathematics: A Linguistic and Philosophical Investigation*. Berlin: Springer.

Gillies, Donald, ed. 1992. *Revolutions in Mathematics*. Oxford: Clarendon.

Goerss, Paul G. 2016. Review of (Hill et al. 2016). *Mathematical Reviews* MR3505179.

Gowers, Timothy. 2012. "Vividness in Mathematics and Narrative." In *Circles Disturbed: The Interplay of Mathematics and Narrative*, edited by Apostolos Doxiadis and Barry Mazur, 211–31. Princeton, NJ: Princeton University Press.

Harris, Michael. 2015. *Mathematics Without Apologies: Portrait of a Problematic Vocation*. Princeton, NJ: Princeton University Press.

Heath, Thomas L. 2006 [1908]. *The Thirteen Books of Euclid's Elements*. New York, NY: Barnes & Noble.

Hill, Michael A., Michael J. Hopkins, and Douglas C. Ravenel. 2010. "The Arf-Kervaire Invariant Problem in Algebraic Topology: Introduction." *Current Developments in Mathematics* 2009: 23–57.

Hill, Michael A., Michael J. Hopkins, and Douglas C. Ravenel. 2016. "On the Nonexistence of Elements of Kervaire Invariant One." *Annals of Mathematics* 184: 1–262.

Howell, P. A. 1985. "The Greek Experience and Aristotle's Analysis of Revolution." In *Revolution: A History of the Idea*, edited by David Close and Carl Bridge, 15–31. Beckenham: Croom Helm.

Jaffe, Arthur, and Frank Quinn. 1993. "'Theoretical Mathematics': Toward a Cultural Synthesis of Mathematics and Theoretical Physics." *Bulletin of the AMS* 29: 1–13.

Ji, Lizhen, and Athanase Papadopoulos. 2013. "Historical Development of Teichmüller Theory." *Archive for the History of the Exact Sciences* 67: 119–47.

Kindi, Vasso. 2015. "The Role of Evidence in Judging Kuhn's Model: On the Mizrahi, Patton, Marcum Exchange." *Social Epistemology Review and Reply Collective* 4(11): 25–33.

Kitcher, Philip. 1993. *The Advancement of Science: Science Without Legend, Objectivity Without Illusions*. Oxford: Oxford University Press.

Klarreich, Erica. 2011. "Mathematicians Solve 45-Year-Old Kervaire Invariant Puzzle." In *The Best Writing on Mathematics 2010*, edited by Mircea Pitici, 373–76. Princeton, NJ: Princeton University Press.

Kuhn, Thomas S. 1970. "Reflections on My Critics." In *Criticism and the Growth of Knowledge*, edited by Imre Lakatos and Alan Musgrave, 231–78. Cambridge: Cambridge University Press.

Kuhn, Thomas S. 1970 [1962]. *The Structure of Scientific Revolutions*, second edition. Chicago: Chicago University Press.

Kuhn, Thomas S. 1974. "Second Thoughts on Paradigms." In *The Structure of Scientific Theories*, edited by Frederick Suppe, 459–82. Urbana, IL: University of Illinois Press.

Kuhn, Thomas S. 2000 [1983]. "Commensurability, Comparability, Communicability." In *The Road since Structure*, edited by James Conant and John Haugeland, 33–57. Chicago: University of Chicago Press.

Lakatos, Imre. 1976. *Proofs and Refutations: The Logic of Mathematical Discovery*. Edited by John Worrall and Elie Zahar. Cambridge: Cambridge University Press.

Lang, Serge. 1994. "Mordell's Review, Siegel's Letter to Mordell, Diophantine Geometry, and 20th Century Mathematics." *Mitteilungen der DMV* 2: 20–31.

Langlands, Robert P. 2013. "Is There Beauty in Mathematical Theories?" In *The Many Faces of Beauty*, edited by Vittorio Hösle, 23–78. Notre Dame, IN: University of Notre Dame Press.

Marcum, James A. 2015. "What's the Support for Kuhn's Incommensurability Thesis? A Response to Mizrahi and Patton." *Social Epistemology Review and Reply Collective* 4(9): 51–62.

Mazur, Barry. 1997. "Conjecture." *Synthese* 111: 197–210.

Mizrahi, Moti. 2015a. "Kuhn's Incommensurability Thesis: What's the Argument?" *Social Epistemology* 29: 361–78.

Mizrahi, Moti. 2015b. "A Reply to James Marcum's 'What's the Support for Kuhn's Incommensurability Thesis?'" *Social Epistemology Review and Reply Collective* 4(11): 21–24.

Mizrahi, Moti. 2015c. "A Reply to Patton's 'Incommensurability and the Bonfire of the Meta-Theories.'" *Social Epistemology Review and Reply Collective* 4(10): 51–53.

Mochizuki, Shinichi. 2013. "On the Verification of Inter-Universal Teichmüller Theory: A Progress Report (as of December 2013)." Online at www.kurims.kyoto-u.ac.jp/~motizuki/IUTeich%20Verification%20Report%202013-12.pdf.

Mochizuki, Shinichi. 2014. "On the Verification of Inter-Universal Teichmüller Theory: A Progress Report (as of December 2014)." Online at www.kurims.kyoto-u.ac.jp/~motizuki/IUTeich%20Verification%20Report%202014-12.pdf.

Otero, Mario H. 1996. "Case Studies as Paradigmatic Exemplars in the Historiography of Mathematics: Inconvenience of a Unified Theory of Radical Change in Mathematics." In *Paradigms and Mathematics*, edited by Elena Ausejo and Mariano Hormigón, 193–200. Madrid: Siglo XXI Editores.

Patton, Lydia. 2015. "Incommensurability and the Bonfire of the Meta-Theories: Response to Mizrahi." *Social Epistemology Review and Reply Collective* 4(7): 51–58.

Pourciau, Bruce. 2000. "Intuitionism as a (Failed) Kuhnian Revolution in Mathematics." *Studies in History and Philosophy of Science* 31: 297–329.

Psillos, Stathis. 1996. "Scientific Realism and the 'Pessimistic Induction.'" *Philosophy of Science* 63: S306–14.

Snaith, Victor P. 2009. *Stable Homotopy around the Arf-Kervaire Invariant*. Basel: Birkhäuser.

Tanswell, Fenner S. 2016. "Proof, Rigour & Informality: A Virtue Account of Mathematical Knowledge." Ph.D. diss., St. Andrews: University of St. Andrews.

Toulmin, Stephen. 1970. "Does the Distinction Between Normal and Revolutionary Science Hold Water?" In *Criticism and the Growth of Knowledge*, edited by Imre Lakatos and Alan Musgrave, 39–47. Cambridge: Cambridge University Press.

Votsis, Ioannis. 2011. "Structural Realism: Continuity and Its Limits." In *Scientific Structuralism*, edited by Peter Bokulich and Alisa Bokulich, 105–17. Dordrecht: Springer.

Zalamea, Fernando. 2012. *Synthetic Philosophy of Contemporary Mathematics*. Falmouth: Urbanomic.

Chapter 8

Revolution or Evolution in Science?

A Role for the Incommensurability Thesis?

James A. Marcum

Revolution is a term that is often abused by historians and philosophers of science in demarcating new science from old and in defending the former from the latter. Thomas Kuhn is certainly guilty of this abuse in his historical philosophy of science and the role incommensurability plays in explicating scientific revolutions. However, later in his career, Kuhn rejected the idea of revolution for that of evolution, and he redefined the role of incommensurability to support his new image of science and conception of scientific change. The change for Kuhn is from larger revolutionary rifts or paradigm shifts to smaller evolutionary specialization or speciation; and, the role of incommensurability swings from accounting for revolutionary rifts to evolutionary speciation. In this chapter, Kuhn's evolutionary image of science is analyzed and critiqued, and a revised version of it is proposed, especially in terms of justifying the role of incommensurability in it. The chapter concludes by exploring the implications of a revised Kuhnian evolutionary image of science for contemporary philosophy of science's pluralistic and perspectival stance.

1. INTRODUCTION

Kuhn's image of science is generally articulated, by either friend or foe, in terms of *The Structure of Scientific Revolutions* (*Structure*)—particularly its first edition, which appeared in 1962 (Hoyningen-Huene 1993, Sharrock and Read 2002). And that image consists of scientific revolutions or paradigm shifts that punctuate normal or paradigmatic science, as figure 8.1 shows.

Moreover, Kuhn proposed a strong form of the incommensurability thesis to support this image of science (Gattei 2008). In other words, for a scientific

Figure 8.1. Kuhn's historical image of science and its progress. Normal science is paradigmatic, according to Kuhn (1962), and the paradigm guides scientific practice and progress until crucial anomalies arise that result in a breakdown of normal science into extraordinary science in which new paradigms are proposed. If the scientific community adopts a new paradigm, then it undergoes a paradigm shift or scientific revolution and a new normal science ensues.

revolution to occur, real and substantial differences must exist between the old and new paradigms such that the two paradigms share little, if anything, in common. However, using the first edition of *Structure* to capture Kuhn's mature image of science is problematic, since Kuhn issued a second edition of it almost a decade later in 1970. In the "Postscript—1969" to that edition, he made significant changes to his image of science, especially with respect to the notion of paradigm. Furthermore, Kuhn (1992) continued to revise his image of science until he himself rejected his original revolutionary philosophy of science for an evolutionary philosophy of science. And, he revised the notion of incommensurability and its role in his new philosophy or image of science (Kuukkanen 2012, Marcum 2015a).

In this chapter, I propose and defend the thesis that Kuhn's later evolutionary image of science and its revised incommensurability thesis—rather than his original revolutionary or historical image and its strong incommensurability thesis—should be the image of science and the notion of incommensurability contemporary philosophers of science engage, in order to discuss the relevance of Kuhn for their discipline's future. I argue that Kuhn's evolutionary philosophy of science and revised incommensurability thesis have important implications and consequences particularly for contemporary philosophy of science, especially given its pluralistic or perspectival stance (Kellert et al. 2006, Giere 2010). To develop and defend this thesis, I begin with a brief narrative that situates Kuhn's original philosophy or image of science and incommensurability thesis in the context of logical positivism and empiricism, as well as falsificationism, and then discuss his new evolutionary image and revised incommensurability thesis. Moreover, a historical case study consisting of the emergence of several microbiological specialties, including bacteriology, virology, and retrovirology, is used to not only illustrate but also critique and revise Kuhn's evolutionary philosophy of science and its incommensurability thesis. I conclude by discussing the implications of a revised Kuhnian evolutionary image of science and its incommensurability thesis for contemporary philosophy of science and its pluralistic and perspectival stance.

2. KUHN'S HISTORICAL PHILOSOPHY OF SCIENCE

Kuhn opens *Structure* with the following well-known proclamation: "History, if viewed as a repository for more than anecdote or chronology, could produce a decisive transformation in the image of science by which we are now possessed" (1962, 1). What was the image of science that Kuhn believed this revisionist history was going to transform? Almost three decades later, in the 1991 Rothschild lecture, Kuhn (1992) summarily answered the question by claiming that the prevailing image of science was at best a caricature of the actual procedures and methods scientists use to investigate and understand the natural world. He likened the image to a "tourist brochure" for describing a country's culture. The philosophical proponents of this image of science were the logical positivists and empiricists, including falsificationists, who envisioned science as a march toward objective and universal facts and truth. In the Rothschild lecture, Kuhn provided the following Polaroid of the traditional or older image of science.

> Science proceeds from facts given by observations . . . . These facts . . . are prior to the scientific laws and theories for which they provide the foundation, and which are themselves, in turn, the basis for explanations of natural phenomena . . . . Unlike the facts on which they are based, these laws, theories, and explanations are not simply given. To find them one must interpret the facts . . . . And interpretation is a human process, by no means the same for all . . . . But, again, observed facts were said to provide a court of final appeal . . . this was the method by which scientists discovered true generalizations about and explanations for phenomena. Or if not exactly true, at least approximations to the truth (1992, 5).

This distorted image of science, as Kuhn contended, gave rise to multiple difficulties, such as observations and facts not being as objective or universal as traditionally believed but imbued with prior values, beliefs, and empirical expectations. As Kuhn concluded, this image had little to do with actual scientific procedures and practices. The remedy was an accurate image or portrayal of science through meticulous and unalloyed narration of its history, especially in terms of the "personal factors" animating scientific practice (Kuhn 1992, 7).

For Kuhn (1962, 1992), the traditional image of science was relatively static compared to science's historical record in which scientific practice was much more dynamic. In fact, such practice was so dynamic—especially in terms of disputes among scientists—that he compared it to a "cat fight" (Kuhn 1992, 6). Science and its practice, then, are not the orderly, logical enterprise as the traditional image portrayed them. Rather, as evident from Kuhn's paradigm concept, multiple elements make up scientific practice. As he articulated the notion of science's dynamism in the "Postscript—1969" and then in the 1991 Rothschild lecture, that notion entails not only observations, facts, theories, hypotheses, and other analytic or objective elements but also nonanalytic or subjective elements such as metaphysical assumptions, values, and personal tastes and interests. Science and its practice, as revealed through an unsullied historical narrative, are not reducible simply to logical and empirical procedures, accompanied by the slow accumulation of scientific facts and knowledge. Rather, scientific advance often involves wholesale breaks or revolutions in not only how scientists go about investigating the world but also what constitutes the world they are investigating. Kuhn (1992) later called this philosophical approach to science as the historical philosophy of science.

An important component of Kuhn's historical philosophy of science is the incommensurability thesis. For the traditional image of science, science's progress and growth were the result of the slow incremental accumulation of facts. But a closer reading of the historical record, claimed Kuhn (1962, 1992), revealed that such progress and growth represented a distortion. Rather, what the historical record depicted was generally a sudden jump in scientific progress in that two competing paradigms, especially an older, established paradigm, often overlapped little, if at all, with a newer competitor paradigm pertaining to facts, observations, terminology, and so on. After all, the competitor paradigm solved the anomalies that led members of a scientific community to question whether the current paradigm could continue to guide normal scientific practice. In other words, the march of science and its progress were not due to the slow accumulation of facts assembled empirically and logically to provide a true picture of the world, but rather scientific change represented a rapid jolt or revolution with respect to an entirely dissimilar way of looking at and investigating the world and understanding it. This radical difference between this old and new image of science was the basis for the incommensurability thesis. Briefly, the stronger version of the thesis states that two competing paradigms are completely dissimilar from each other to such an extent that they do not share even the same view of the world (Horwich 1993).[1] The classic example is the Copernican Revolution in which the sun was no longer a planet roaming the earthly sky but the center of the universe, while the earth was dethroned to a wandering planet, around the solar sky (Kuhn 1957).

3. KUHN'S EVOLUTIONARY PHILOSOPHY OF SCIENCE

Later in his career, however, Kuhn himself was to undergo a revolutionary change in his image of science, from a historical to an evolutionary one. In the 1991 Rothschild lecture, he acknowledged that there were serious problems that emerged with historical philosophy of science. Basically, he claimed that the historiographic revolution in science studies undercut two chief pillars supporting the traditional image of science, without offering anything to replace them. The two pillars were,

> first, that facts are prior to and independent of the beliefs for which they are said to supply the evidence, and, second, that what emerges from the practice of science are truths, probable truths, or approximations of the truth about a mind- and culture-independent external world (1992, 18).

As Kuhn laments, in the aftermath of the historiographic revolution, advocates of historical philosophy of science tried in vain to eradicate any trace of the traditional image of science or to reinstall a chastened version of it. And, professional and academic life became burdensome for them. He offered an alternative solution to the problems associated with the historiographic revolution—evolutionary philosophy of science.

Kuhn makes a shift in what he considers to be the agenda for historians and philosophers of science. For historical philosophy of science, he claims, the agenda was to explain major changes in scientific practice and knowledge; but this agenda led to the problems associated with historical philosophy of science undermining the two pillars of traditional philosophy of science. Now, according to Kuhn, the agenda should be the explanation of "small incremental *changes* of belief" (1992, 11). Rather than the upheaval of world-shattering revolutions in the wake of science's advancement, scientific progress should be viewed as the gradual proliferation of scientific specialties much akin to biological speciation. Briefly, for Kuhn, science progresses by gradual emergence of a specialty's practice and knowledge. As the members of an established specialty practice their trade, a new specialty arises because of novel and anomalous results, which requires alterations in the established specialty's lexicon—the substitute for historical philosophy of science's notion of paradigm. Instead of the unity of science, Kuhn now envisions science pluralistically—which has important implications for his contribution to contemporary philosophy of science, as discussed in the concluding section.

With the shift from a historical to an evolutionary philosophy of science, Kuhn (1992) also changed both the notion of and the role for the incommensurability thesis (Marcum 2015b). Whereas gradualism represents the rate of change for his evolutionary philosophy of science, incommensurability represents the mechanism for specialization. And, he now defines the incommensurability thesis in terms of alterations in the lexical taxonomy or structure of a scientific specialty—what Howard Sankey (1998) calls "taxonomic incommensurability." Instead of no common meaning, as Kuhn originally defined the notion of incommensurability, he now defines the notion as no common taxonomy. In other words, as science advances in terms of increased specialization, the lexical taxonomy of referring terms to objects changes dramatically with addition or deletion of concepts and terms within the original lexicon. Particularly, new concepts and terms are introduced that result in the appearance of a new lexicon with a significantly different mapping of objects and the terms referring to them.

Kuhn also changes incommensurability's role with respect to paradigm shifts or scientific revolutions for historical philosophy of science to a mechanism for isolating lexicons of different scientific specialties, so that a new specialty can split off from the established or parent specialty and evolve into an independent specialty. However, as I argue in further section for a revised version of Kuhn's evolutionary philosophy of science, the emergence of a specialty—like retrovirology—might not involve radical taxonomic incommensurability with a parent specialty in that the new specialty might not change all the terms and concepts within an established lexicon but simply add novel terms and concepts to it. In the next section, the emergence of several specialties in the microbiological sciences is reconstructed not only to assess Kuhn's evolutionary philosophy of science but also to revise it.

4. FROM BACTERIOLOGY TO VIROLOGY TO RETROVIROLOGY

The contemporary history of microbiological specialties, in terms of the emergence or evolution of bacteriology, virology, and retrovirology, serves to illustrate (and to critique) Kuhn's evolutionary philosophy of science— with respect to its strengths and weakness—and this case history provides an example for discussing, in the final section, the impact and implications of his new image of science for contemporary philosophy of science. Although the case history of these microbiological specialties represent only one example to support an evolutionary image of science, the expectation is that other case histories would also support this image. This expectation is not unreasonable, since examination of the historical record for scientific progress supports a diversity of times or tempos and means or modes. For example, the neurosciences have evolved in terms of the emergence of new specialties such as cognitive neuroscience (Boden 2006) and synaptic physiology (Bennett 2001), which exhibit significant differences with respect to their evolutionary pace and mechanism.

The appearance of bacteriology, in particular, represents specialization in terms of infectious pathology (Lederberg 2000). Contemporary infectious pathology begins with the germ theory of disease, in the context of the debate between the miasma and contagion theories of disease during the latter part of the nineteenth century (Magner 2009, Mitchell 2012). Briefly, the miasma theory explained the occurrence of epidemics and infectious diseases in terms of dirty or noxious air, while the contagion theory claimed that these diseases were spread through direct contact with an infectious agent. Although the exact nature of the causative or infectious agent was unknown, both theories targeted decaying and putrid material as the source of the agent.

The miasma and contagion theories had very different lexicons to account for the causative agents and sources of epidemics and infectious diseases, such as cholera and yellow fever. Before a committee of Parliament on March 5, 1855, for example, John Snow's testimony against the Nuisances Removal and Diseases Prevention Act illustrates the lexical differences between the two theories (Lilienfeld 2000). Snow's interlocutor, Sir Benjamin Hall—who was an advocate of the miasma theory—asked Snow whether inhaling or smelling "offensive effluvia" would be injurious to a person's health. Snow answered he believed it would not; rather, in response to further questions about the transmission of infectious disease, he defended the belief that con- tact with an infectious agent was necessary. Moreover, when pressed with examples of disease resulting from inhalation of noxious air from decaying material known to cause infectious disease, Snow claimed that the resulting illness was a "coincidence" of the concentration of the noxious gases but was not directly caused by them. Snow's lexicon simply did not contain notions such as offensive effluvia for explaining the cause of infectious disease; rather, it had terms such as gases and contagion.

By the end of the nineteenth century, however, the germ theory of disease eclipsed both the miasma and contagion theories. The germ theory repre- sented a major change or shift in understanding and explaining epidemics and infectious diseases, compared to the miasma and contagion theories. Indeed, John Waller (2002) refers to it as the "germ revolution." And, the theory's proponents introduced new referring terms for novel entities—such as bac- teria. Bacteria or germs are not simply a referential extension of or addition to the decaying material of either the miasma or the contagion theory; rather, they represent an ontologically distinct class of entities with an ability to cause disease (Pelting 1993).[2] Moreover, a large part of the justification of the germ theory was technological in terms of the isolation and visualization of bacteria. The role of the microscope in visualizing such organisms is well documented historically (Bradbury 1967). However, the method for isolating them was equally important. That method utilized a filtration technique in which a sample was passed over a selective filter that retained the bacteria (Magner 2009). The combined result of these research techniques into infectious disease was the scientific specialty of bacteriology (Bulloch 1938, Foster 1970).

During the nineteenth century, virology—as a separate specialty—emerged from bacteriology (Hughes 1977, Waterson and Wilkinson 1978). Compared to the establishment of the germ theory and bacteriology, which took over a century, virology occurred relatively rapidly and dramatically. As for bacteriology, filtering techniques were also critical for virology's emergence and development. For example, Dmitri Ivanovski, while investigating tobacco mosaic disease during the late nineteenth century, discovered that the infectious agent passed through a filter that retained bacteria (Creager 2002, Grafe 2012). The issue, however, was whether the infective agent was a chemical toxin or resembled living organisms like bacteria. Subsequent research demonstrated that the agent was similar to a living organism rather than to a chemical toxin, but it did not exhibit completely the structure of living organisms, not even bacteria. Moreover, there were significant differences in the chemical makeup of viruses, especially with respect to their genomic material. For prokaryotes and eukaryotes, the genome is composed of DNA exclusively; however, some viruses have an RNA genome (Strauss and Strauss 2008, Wassenaar 2012, Cooper and Hausman 2013). Thus, the virology lexicon was significantly different from the bacteriology lexicon, especially in terms of the genome's genetic composition.[3]

But even the virology lexicon was going to be significantly altered in the early part of the twentieth century.[4] Peyton Rous (1911), in particular, published findings suggesting that a virus was responsible for causing cancer in chickens. Although the scientific community was initially skeptical about the findings, by mid-twentieth century, Richard Shope—a younger colleague of Rous—had convinced the community of Rous' findings, and the virus was named the Rous sarcoma virus or RSV (Marcum 2001). RSV proved to be more unusual than simply causing cancer in chickens. Its genome was composed of RNA, although that was not entirely novel since other viruses' genomes were known to be composed of RNA. What was unusual was the mechanism by which the genome propagated itself. For most RNA viruses, the genome is simply transcribed into additional RNA, which is then assembled along with other components as the virus. For RSV, however, its RNA genome is first transcribed by reverse transcriptase into DNA—as the DNA provirus—and then incorporated into the host genome (Marcum 2002). Only later is the DNA provirus retranscribed into RNA and then assembled with other components as the virus. Thus, a new subspecialty—retrovirology—branched off from virology in the latter part of the twentieth century. Moreover, retrovirology eventually led to the discovery of the cancer-causing genes, oncogenes (Vogt 1997).

5. REVISING KUHN'S EVOLUTIONARY IMAGE OF SCIENCE AND INCOMMENSURABILITY

Kuhn's evolutionary philosophy of science does provide a general philosophical approach for reconstructing and analysing the emergence and evolution of the microbiological specialties of bacteriology, virology, and retrovirology. In Kuhnian terms, each of these specialties had a particular lexicon, with certain incommensurable terms and concepts, for describing and explaining natural phenomena associated with epidemics and infectious diseases, which eventually included other specialties like oncology. Moreover, these lexicons served to isolate each of the specialties from one another and to allow them the space, both physically and conceptually, to evolve independently. For example, virology introduced a completely novel class of entities—viruses—for determining the etiology of certain infectious disease, while retrovirology expanded the type of the diseases from infection to cancer. Thus, Kuhn's evolutionary philosophy of science with its notion of lexical incommensurability allows philosophers and historians of science to map the relationships of the different specialties and how these relationships contribute to the growth of scientific knowledge. However, as the case history of bacteriology, virology, and retrovirology demonstrates, Kuhn's evolutionary philosophy of science does have its limitations. Specifically, Kuhn accounts for the evolution of science with respect to a single tempo and mode of evolutionary change—the tempo is Darwinian gradualism and the mode speciation. But, evolutionary tempo and mode are more diverse than these two options.

The twentieth-century palaeontologist, George Gaylord Simpson (1944), provides a useful taxonomy of evolutionary tempos and modes for revising Kuhn's evolutionary philosophy of science (Damuth 2001, Kutschera and Niklas 2004). Briefly, Simpson identified three types of tempos. The first is bradytelic, which represents one end on the tempo spectrum in that it is the slowest rate of evolutionary change. The next tempo, which he called tachytelic, represents the other end of the spectrum and is the fastest rate of evolutionary change. The final tempo represents an intermediate position between the two ends. He considers it the standard or typical rate of evolutionary change and calls it horotelic.

Simpson also identified three modes of evolution, which parallel loosely the three tempos—although each of the tempos can be associated to some degree with any of the modes. The first mode is phyletic evolution in which a whole population gradually shifts over time to a new taxonomic entity. It is generally associated with a bradytelic tempo and often referred to as phyletic gradualism. The next mode is quantum evolution and is mostly associated with a tachytelic mode in which a new taxonomic entity appears rather suddenly. The final mode is speciation and is generally associated with a horotelic tempo in which a new taxonomic entity branches off from a parent taxonomic entity.

Simpson's taxonomy of evolutionary tempos and modes can be used to analyze the emergence or evolution of the microbiological specialties of bacteriology, virology, and retrovirology, as shown in table 8.1.

The tempo and mode for bacteriology's appearance were bradytelic and phyletic evolution, respectively. The tempo was bradytelic since it took over a century to unfold and was intimately associated with the debate over the miasma and contagion theories of epidemics and infectious diseases, as well as with the establishment of the germ theory of disease. The mode was phyletic evolution since it represented a shift along the epistemic path of understanding the etiology of infectious diseases. The emergence of virology, however, exhibited a tachytelic tempo and a mode of quantum evolution since it emerged relatively quickly and dramatically in comparison to bacteriology and represented a major leap in the nature of a novel and unanticipated class of infectious agents. Finally, the appearance of retrovirology represented a mode of speciation, splitting off from virology, at a horotelic tempo.

The revised Kuhnian evolutionary philosophy of science takes into consideration the distinct tempos and modes as introduced by Simpson, and as illustrated in the earlier discussed historical case study, in order to afford a more comprehensive approach to science's evolution. Thus, it provides a robust and diverse approach for describing and explaining the nature of science and the growth of scientific knowledge. As for the nature of science, the proposed evolutionary philosophy or image of science does not limit science to a set of methodological procedures or principles, especially logical rules or algorithms, which are fixed and unalterable. Rather, as evident from the history of science, there has been the emergence of important methodological principles during the evolution of science, especially since the scientific revolution (Gower 1997).[5] For example, the development of

Table 8.1. Tempo and mode for the evolution of microbiological specialties of bacteriology, virology, and retrovirology (see text for details)

Scientific Specialty	Tempo	Mode
Bacteriology	Bradytelic	Phyletic
Virology	Tachytelic	Quantal
Retrovirology	Horotelic	Speciation

controlled experiments has been crucial for investigating natural phenomena and establishing the veracity of scientific claims (Boring 1954, Mayo 1996). Also, there has been technological innovations that are significant in the emergence of scientific specialties. As mentioned earlier in the chapter, the development of the microscope, along with the electron microscope and various staining protocols, were critical for the evolution of the various microbiology specialties.

For the growth of scientific knowledge, the revised Kuhnian evolutionary philosophy of science again affords a variety of approaches for describing how such knowledge grows and is incorporated in a specialty's lexicon. Importantly, the growth need not be in terms of rapid or gradual tempos or rates only; but, it can exhibit a variety of rates in between these two ends of the spectrum—and these intermediate rates might be the general or common rates of growth. In other words, both bradytelic and tachytelic growth do occur, as illustrated by bacteriology and virology, respectively, but like biological evolution, most growth in terms of conceptual evolution is horotelic, as illustrated by the emergence of retrovirology. Moreover, the evolution of science and scientific knowledge involves not only speciation but also phylogenesis. Not only do new scientific specialties, similar to biological species, emerge but also do whole new areas of science, similar to phylogenesis, as illustrated with the emergence of bacteriology. Another important example of a phyletic mode of evolution along with a horotelic tempo is the emergence of systems biology, especially in terms of the its holistic metaphysical foundations compared to the reductionist metaphysics of molecular biology (Marcum 2008).

What is the advantage, then, of the revised Kuhnian evolutionary philosophy of science, as compared to Kuhn's original version of it? There are as least two distinct advantages. First, the revised version provides a richer and more textured approach to analyzing the emergence of scientific specialties than Kuhn's original version, especially in terms of tempo and mode. Relying only on Darwinian gradualism, as Kuhn does, can be a rather myopic perspective of science, which fails to envision other possible tempos and modes for the evolution of scientific practice and knowledge. Also, relying only on such gradualism suffers from the traditional philosophical agenda of attempting to unify the sciences, that is, all sciences evolve gradually and steadily from a common ancestor—physics. This urge to unify the sciences ran deep in the subconscious of a previous generation of philosophers of science—and Kuhn was not immune from it. This leads to the second advantage: the revised evolutionary philosophy of science promotes the diversity of evolutionary tempos and modes and thereby assists the advancement of the pluralistic and perspectival stance of contemporary philosophy of science—to which we turn to now.

6. CONTEMPORARY PHILOSOPHY OF SCIENCE

In contrast to traditional twentieth century philosophy of science with its "unity of science" agenda, contemporary philosophy of science is pluralistic and perspectival in its stance and approach to examining and explicating science (Kellert et al. 2006; Giere 2010). This pluralism and perspectivism can be divided into two major categories. The first is the philosophy of the sciences in which specific sciences often have different philosophical issues facing them. For example, the philosophical issues confronting evolutionary biologists concerning the notion of species is very different from chemists concerning the notion of atomic structure. In a pluralistic or perspectival approach, no one science dominates or serves as the model for the other sciences when philosophically investigating a particular science. Specifically, this means that physics, traditionally considered to be the paradigmatic science, if you will, no longer functions as the model or standard for investigating and explaining the nature of science per se or specific nonphysical sciences. In other words, physics does not serve as the perspective or standard from which the other sciences, like chemistry or biology, are viewed and evaluated in terms of their scientific status.

The second category pertains to various philosophical approaches besides the traditional twentieth century logico-analytical approach, which again served as the paradigmatic approach to addressing philosophical issues of the sciences—particularly the physical sciences. For example, phenomenological or continental philosophy provides a complementary and often challenging approach to philosophical investigation of science, which the Anglophone analytic philosophical tradition has generally marginalized (Gutting 2005). Martin Heidegger (1966, 45–47), for instance, engages in an insightful analysis of how we have become thought-less or thought-poor by relying solely on calculative thinking, which only calculates possibilities of events but not their meanings, which Heidegger claims requires meditative thinking. Several other philosophical traditions, such as pragmatism (Almeder 2007; Kitcher 2013) and feminism (Keller and Longino 1996; Tuana 1989), are also utilized to address issues in science. Gone, then, are the days when a single philosophical approach or tradition—particularly the logico-analytical tradition—provides the resources for doing philosophy of science, especially in terms of unifying the sciences or reducing them to physics.

Given the state of contemporary philosophy of science, does evolutionary philosophy of science—and particularly Kuhn's version of it—represent a candidate for general philosophy of science (Psillos 2012; 2016)? In other words, is evolutionary philosophy of science the next major philosophy for science to replace historical philosophy of science or other philosophical approaches toward science studies? Or is it simply one approach among many, like phenomenology, analytic philosophy, historical philosophy of science, and so on? My thesis is that evolutionary philosophy of science can embrace contemporary philosophy of science's pluralism and perspectivism, but a robust rather than a fragile formulation needs to be developed. In other words, the robust formulation should accommodate the various natural and even social sciences from different philosophical perspectives.

As discussed in the preceding section, the problem with Kuhn's version of evolutionary philosophy of science, and with most—if not all—versions of evolutionary philosophy of science, is that the proponents are generally committed to one particular tempo and mode of evolutionary change—the tempo is Darwinian gradualism and the mode speciation (Marcum 2015a). However, evolutionary tempo and mode are more diverse than these two options. In other words, Kuhn's original version of evolutionary philosophy of science is rather fragile in the sense that it cannot account for the multiple ways by which science is practiced and scientific knowledge evolves and grows. As the historical case study demonstrates, the evolution of the microbiological specialties involved a host of tempos and modes, and hence requires a robust image to account for their practice and growth of knowledge.

Moreover, the evolutionary image of science is organic and not simply logico-analytical or even social. The evolutionary image captures the dynamism and complexity involved in scientific practice and the growth of scientific knowledge, especially with respect to the proliferation of scientific specialties and subspecialties. Although the unity of science in terms of reducing the sciences to physics is no longer an agenda item for contemporary philosophers of science, there is a type of unity that emerges with evolutionary philosophy of science, based on the notion of common descent in evolutionary biology. Specifically, the relationship of the diverse specialties can be mapped in relation to one another in order to provide conceptual diagrams of particular scientific disciplines. For example, both infection and cancer are related to one another in the sense that they involve a compromise in an organism's immune system, as figure 8.2 shows. Moreover, understanding these relationships can assist researchers in developing effective treatments for patients, especially cancer patients (Munn 2016).

Finally, I would like to address briefly a recent exchange over whether there is support for Kuhn's incommensurability thesis. Moti Mizrahi (2015a) argued that there are neither valid deductive nor strong inductive arguments for Kuhn's thesis of taxonomic incommensurability. In response to Mizrahi, Lydia Patton (2015) claimed that abductive arguments or inferences to the best explanation are generally the preferred form of argumentation in the history and philosophy of science. In response to both, I suggested that Kuhn's incommensurability thesis could be supported by the historian's experience of engaging a science's history—particularly when trying to make sense of texts that at first seem at odds with current theories (Marcum 2015c).[6] In light

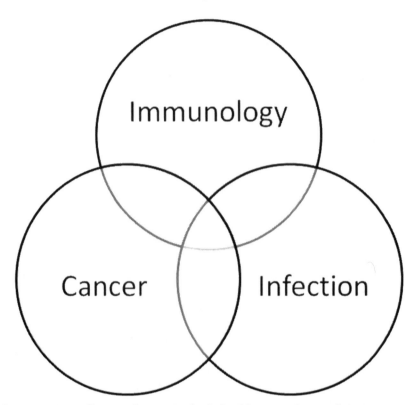

Figure 8.2. Venn diagram of conceptual relationship among immunology, cancer, and infection. The proposed evolutionary philosophy of science provides a type of unity not in terms of reducing scientific specialties to a common discipline like physics or chemistry—as for the traditional analytic approach—but with respect to the relationship among specialties particularly in terms of their conceptual content. In the example, each specialty can intersect with another or remain partially independent.

of the forgoing discussion on evolutionary philosophy of science and the role incommensurability plays in it, I contend that this role provides additional support for the notion of incommensurability. In other words, incommensurability is an essential component of an evolutionary image of science and the growth of scientific knowledge. Just as environmental isolation is essential for biological evolution, so incommensurability is critical for conceptual evolution within the sciences.

7. CONCLUSION

In sum, Kuhn's evolutionary image of science does provide a fecund starting point for developing a philosophy of science that addresses the evolving nature of science in which incommensurability plays a significant role in scientific progress with respect to specialization within the sciences. Specialization, for Kuhn, is the key to understanding the world better, not so much in terms of absolute or universal truth, but with respect to the details that make up the world through carving nature closer and more precisely at its evolutionary joints. However, his evolutionary philosophy of science is not robust enough to analyze the development of specialization for the sciences at large. The revised Kuhnian evolutionary philosophy of science proposed in this chapter addresses that problem in terms of robustness, by introducing a variety of modes and tempos and by illustrating it with a case history from the microbiological specialties of bacteriology, virology, and retrovirology. Moreover, this revised image of science not only respects contemporary philosophy's pluralistic and perspectival stance but also offers a type of unity among the sciences, not in terms of reducing them to one science, but rather with respect to mapping the conceptual relationships among them. Obviously, there is much more work to do in terms of providing a complete evolutionary philosophy of science that might bring some degree of consensus to contemporary philosophy of science.

NOTES

1 Kuhn (1982) acknowledged a weaker version of the incommensurability thesis in which two competing paradigms are not so radically dissimilar but do share to some extent common terms and concepts.

2 Although a semantic change does not necessarily entail taxonomic incommensurability (Mizrahi 2015a), it does not exclude it. In this historical case, the introduction of novel entities account for conducting science within a world different from that in terms of the miasma or contagion theory.

3 What made the virus lexicon so different in terms of an RNA genome is that initially the biological community thought that genetic information flowed from DNA to RNA to protein (The Central Dogma) and not in reverse, although there was some debate about whether information could flow from RNA to DNA with respect to viruses (Marcum 2001, 2002). But for prokaryotes and eukaryotes, the flow was definitely thought to be unidirectional.

4 Whereas the virology and bacteriology lexicons are significantly different vis-à-vis incommensurable terms and concepts, the virology and retrovirology lexicons differ only in terms of the addition of novel terms and concepts for retroviruses.

5 It is important to note that although taxonomic incommensurability is concerned chiefly with theoretical terms and concepts, such incommensurability is based on or intimately connected to the methodology and techniques used to investigate nature or to practice science. For example, big data–driven science and hypothesis-driven science take very different, if not incommensurable, approaches toward investigating natural or social phenomena (Mazzocchi 2015).

6 Mizrahi (2015b) responded to Marcum (2015c), claiming that the historian's experience of incommensurability in historical texts begs the question of taxonomic incommensurability since such experience assumes incommensurability. But, I would contend that the experience of incommensurable texts does not beg the question but rather functions to provide meaning to the experience. Indeed, Mizrahi concludes his comments claiming that our differing interpretations of a historical case study concerning Galen and Harvey supports his claim that "there is no strong inductive support (in the form of inductive generalizations from case histories) for TI [taxonomic incommensurability]" (2015b, 23). Although such differences between us in interpreting the case study might not lend inductive support to taxonomic incommensurability logically, from the experience of the person reconstructing the historical case study such differences do support incommensurability since we both reconstruct the case study in terms of its meaning for supporting or not supporting taxonomic incommensurability. In other words, I have to reconstruct Mizrahi's critique of my comments in order to make sense of them. In Kuhnian terms, Mizrahi and I "do *in some sense* live in different worlds" (1970, 193).

REFERENCES

Almeder, Robert. 2007. "Pragmatism and Philosophy of Science: A Critical Survey." *International Studies in the Philosophy of Science* 21: 171–95.

Bennett, Max R. 2001. *History of Synapse*. Amsterdam: OPA.

Boden, Margaret. 2006. *Mind as Machine: A History of Cognitive Science, Volume 1 & 2*. New York: Oxford University Press.

Boring, Edwin G. 1954. "The Nature and History of Experimental Control." *American Journal of Psychology*: 67: 573–89.

Bradbury, Savile. 1967. *The Evolution of the Microscope*. Oxford: Pergamon Press.

Bulloch, William. 1938. *The History of Bacteriology*. Oxford: Oxford University Press.

Cooper, Geoffrey M., and Robert E. Hausman. 2013. *The Cell: A Molecular Approach*. Sunderland, MA: Sinauer Associates.

Creager, Angela N. H. 2002. *The Life of a Virus: Tobacco Mosaic Virus as an Experimental Model, 1930–1965*. Chicago: University of Chicago Press.

Damuth, John D. 2001. "Evolution: Tempo and Mode." *Encyclopedia of Life Sciences*. Accessed January 5, 2017. doi: 10.1038/npg.els.0001720.

Foster, William D. 1970. *A History of Medical Bacteriology and Immunology*. London: Heinemann.

Gattei, Stefano. 2008. *Thomas Kuhn's "Linguistic Turn" and the Legacy of Logical Empiricism: Incommensurability, Rationality, and the Search for Truth*. New York; Routledge.

Giere, Ronald N. 2010. *Scientific Perspectivism*. Chicago: University of Chicago Press.

Gower, Barry. 1997. *Scientific Method: An Historical and Philosophical Introduction*. New York: Routledge.

Grafe, Alfred. 2012. *A History of Experimental Virology.* New York: Springer.

Gutting, Gary, editor. 2005. *Continental Philosophy of Science.* Oxford: Blackwell.

Heidegger, Martin. 1966. *Discourse on Thinking.* New York: Harper & Row.

Horwich, Paul, editor. 1993. *World Changes. Thomas Kuhn and the Nature of Science.* Cambridge, MA: MIT Press.

Hoyningen-Huene, Paul. 1993. *Reconstructing Scientific Revolutions: Thomas S. Kuhn's Philosophy of Science.* Chicago: University of Chicago Press.

Hughes, Sally Smith. 1977. *The Virus: A History of the Concept.* London: Heinemann.

Keller, Evelyn Fox, and Helen E. Longino, editors. 1996. *Feminism and Science.* New York: Oxford University Press.

Kellert, Stephen H., Helen E. Longino, and C. Kenneth Waters, editors. 2006. *Scientific Pluralism.* Minneapolis: University of Minnesota Press.

Kitcher, Philip. 2013. "Toward a Pragmatist Philosophy of Science." *Theoria* 77: 185–231.

Kuhn, Thomas S. 1957. *The Copernican Revolution: Planetary Astronomy in the Development of Western Thought.* Cambridge, MA: Harvard University Press.

Kuhn, Thomas S. 1962. *The Structure of Scientific Revolutions.* Chicago: University of Chicago Press.

Kuhn, Thomas S. 1970. *The Structure of Scientific Revolutions*, 2nd edition. Chicago: University of Chicago Press.

Kuhn, Thomas S. 1982. "Commensurability, Comparability, Communicability." *PSA: Proceedings of the Biennial Meeting of the Philosophy of Science Association*, vol. 1982, 669–88.

Kuhn, Thomas S. 1992. *The Trouble with Historical Philosophy of Science.* Cambridge, MA: Department of the History of Science, Harvard University.

Kutschera, Ulrich, and Karl J. Niklas. 2004. "The Modern Theory of Biological Evolution: An Expanded Synthesis." *Naturwissenschaften* 91: 255–76.

Kuukkanen, Jouni-Matti. 2012. "Revolution as Evolution: The Concept of Evolution in Kuhn's Philosophy." In *Kuhn's* The Structure of Scientific Revolutions *Revisited*, edited by Vasso Kindi and Theodore Arabatzis, 134–52. New York: Routledge.

Lederberg, Joshua. 2000. "Infectious History." *Science* 288: 287–93.

Lilienfeld, David E. 2000. "John Snow: The First Hired Gun?" *American Journal of Epidemiology* 152: 4–9.

Magner, Lois N. 2009. *A History of Infectious Disease and the Microbial World.* Westport, CT: Praeger.

Marcum, James A. 2001. "The Transformation of Oncology in the Twentieth Century: The Molecularization of Cancer." In *Proceedings of the 37th International Congress on the History of Medicine*, edited by C. R. Burns, Y. V. O'Neill, P. Albou, and J. G. Rigau-Pérez, 41–49. Galveston, TX: University of Texas Medical Branch.

Marcum, James A. 2002. "From Heresy to Dogma in Accounts of Opposition to Howard Temin's DNA Provirus Hypothesis." *History and Philosophy of the Life Sciences* 24: 165–92.

Marcum, James A. 2008. "Does Systems Biology Represent a Kuhnian Paradigm shift?" *New Phytologist* 179: 587–89.

Marcum, James A. 2015a. *Thomas Kuhn's Revolutions: A Historical and an Evolutionary Philosophy of Science?* London: Bloomsbury.

Marcum, James A. 2015b. "The Evolving Notion and Role of Kuhn's Incommensurability Thesis." In *Kuhn's Structure of Scientific Revolutions-50 Years On*, edited by William J. Devlin and Alisa Bokulich, 115–34. New York: Springer.

Marcum, James A. 2015c. "What's the Support for Kuhn's Incommensurability Thesis? A Response to Mizrahi and Patton." *Social Epistemology Review and Reply Collective* 4(9): 51–62.

Mayo, Deborah G. 1996. *Error and the Growth of Experimental Knowledge*. Chicago: University of Chicago Press.

Mazzocchi, Fulvio. 2015. "Could Big Data Be the End of Theory in Science?" *EMBO Reports* e201541001.

Mitchell, Peta. 2012. *Contagious Metaphor*. London: Bloomsbury.

Mizrahi, Moti. 2015a. "Kuhn's Incommensurability Thesis: What's the Argument?" *Social Epistemology* 29: 361–78.

Mizrahi, Moti. 2015b. "A Reply to James Marcum's What's the Support for Kuhn's Incommensurability Thesis?" *Social Epistemology Review and Reply Collective* 4(11): 21–24.

Munn, Lance L. 2016. "Cancer and Inflammation." *WIREs System Biology and Medicine* 9. Accessed February 12, 2017. doi: 10.1002wsbm.1370.

Patton, Lydia. 2015. "Incommensurability and the Bonfire of the Meta-theories: Response to Mizrahi." *Social Epistemology Review and Reply Collective* 4(7): 51–58.

Pelting, Margaret. 1993. "Contagion/Germ Theory/Specificity." In *Companion Encyclopedia of the History of Medicine*, Vol. 1, edited by William F. Bynum and Roy Porter, 309–34. New York: Routledge.

Psillos, Stathis. 2012. "What Is General Philosophy of Science?" *Journal for General Philosophy of Science* 43: 93–103.

Psillos, Stathis. 2016. "Having Science in View: General Philosophy of Science and Its Significance." In *The Oxford Handbook of Philosophy of Science*, edited by Paul Humphreys, 137–62. New York: Oxford University Press.

Rous, Peyton. 1911. "A Sarcoma of the Fowl Transmissible by an Agent Separable from the Tumor Cells." *Journal of Experimental Medicine* 13: 397–411.

Sankey, Howard. 1998. "Taxonomic Incommensurability." *International Studies in the Philosophy of Science* 12: 7–16.

Sharrock, Wes, and Rupert Read. 2002. *Kuhn: Philosopher of Scientific Revolutions*. Malden, MA: Polity.

Strauss, Ellen G., and James H. Strauss. 2008. *Viruses and Human Diseases*, 2nd edition. New York: Elsevier.

Tuana, Nancy, editor. 1989. *Feminism and Science*. Bloomington, IN: Indiana University Press.

Vogt, Peter K. 1997. "Historical Introduction to the General Properties of Retroviruses." In *Retroviruses*, edited by John M. Coffin, Stephen H. Hughes, and Harold E. Varmus, 1–25. New York: Cold Spring Harbor Laboratory Press.

Waller, John. 2002. *The Discovery of the Germ: Twenty Years That Transformed the Way We Think about Disease*. New York: Columbia University Press.

Wassenaar, Trudy M. 2012. *Bacteria: The Benign, the Bad, and the Beautiful*. Hoboken, NJ: John Wiley & Sons.

Waterson, Anthony P., and Lise Wilkinson. 1978. *An Introduction to the History of Virology*. Cambridge: Cambridge University Press.

Part IV

ABANDONING THE KUHNIAN IMAGE OF SCIENCE

Chapter 9

The Biological Metaphors of Scientific Change

Barbara Gabriella Renzi
Giulio Napolitano

In this chapter, we take the philosophical stance that cognitive metaphors shape in fundamental ways not only how scientists communicate but also how they think. Philosophers and scientists constantly employ conceptual metaphors in their explanations, and Kuhn, over the years, makes few attempts at providing an evolutionary metaphor for scientific change. Here we discuss how biological analogies may ultimately be inadequate for the description of scientific change and how this inadequacy affects the soundness of Kuhn's idea of incommensurability.

1. LANGUAGE AND COGNITION

Over the time, theories of metaphor have oscillated between opposite approaches. Is metaphor a linguistic or a cognitive phenomenon? Do metaphors contribute to the creation of new meaning or do they only convey literal interpretations of words? Are metaphors useful, explicative devices or do they introduce confusion and ambiguity? These distinctions were already clearly introduced in the first extended philosophical treatment of metaphor given by Aristotle. The Greek philosopher considered metaphors as potent tools enabling the understanding of hidden truths, by means of transferral of characteristics, analogy (Aristotle 1941, 1475b) and similarity relations (Aristotle 1941, 1459s). He located the transferral at the level of individual words, and this approach continued for a long time, as only in the twentieth century has the metaphoric transfer been situated at the level of sentences, after the realization that the semantic unit is larger than a word (Ricoeur 1977). Aristotle also regarded metaphor as a deviance from the normal usage of language, maintaining that "diction becomes distinguished and non-prosaic by the use of unfamiliar terms, i.e. strange words, metaphors, lengthened forms, and everything that deviates from ordinary modes of speech" (Aristotle 1941, 148a). For him, metaphor is only slightly different from simile[1] (Aristotle 1941, 1406b) and this standpoint is the foundation of the so-called Traditional View (Johnson 1981), in which metaphor is considered valuable only for didactic purposes and for stylistic and rhetorical goals. Metaphors can be translated into literal paraphrases without losing cognitive content, leading to the view that metaphor is an elliptical simile.

The *linguistic* phenomenon approach may be exemplified by the views expressed by Donald Davidson, who affirmed that, as metaphorical statements are clearly identified as such by a receiver, they do not create new meaning. Metaphors only have a literal meaning, they "mean what the words, in their literal interpretation, mean, and nothing more" (Davidson 1978, 32). When a statement such as "a man is an island" is heard or read, anybody will have no doubt that this is a metaphorical statement. The individual word meanings, however, are not changed by the metaphorical collocation.

> I depend on the distinction between what words mean and what they are used to do. I think metaphor belongs exclusively to the domain of use. It is something brought off by imaginative employment of words and sentences and depends entirely on the ordinary meanings of those words and hence on the ordinary meanings of the sentences they comprise (Davidson 1978, 33).

To Davidson, metaphors are only pragmatic devices that lead us to become aware of something that otherwise we would not have noticed.

More recently, the *cognitive* approach has gained significant interest and is now largely adopted, with some scholars claiming that our understanding of the world only occurs as mediated through our metaphors. Metaphor is undeniably omnipresent in our everyday life, because the nature of our ordinary conceptual system is fundamentally metaphorical:

> The locus of metaphor is not in language at all, but in the way we conceptualize one mental domain in terms of another. The general theory of metaphor is given by characterizing such cross-domain mappings. And in the process, everyday abstract concepts like time, states, causation and purpose also turn to be metaphorical (Lakoff 1993, 203).

In the view proposed by George Lakoff and Mark Turner, among others, the relationship between language and cognition is reversed: metaphors are conceptual mappings and are "primarily a matter of thought and action and only derivatively a matter of language" (Martinich 1984). It is the presence of metaphors in our patterns of thought that causes their appearance in our language, not vice versa.

> Nearly always, when we talk about abstract concepts, we choose language drawn from one or another concrete domain. A good example of this is our talk about the mind. Here we use the spatial model to talk about things that are clearly nonspatial in character. We have things "in" our mind, "on" our minds, "in the back corners of" our minds. We "put things out" of our minds, things "pass through" our minds, we "call things to mind," and so on. It is quite possible that our primary method of understanding nonsensory concepts is through analogy with concrete experiential situations (Rumelhart 1993, 71).

In this view, there is a crucial, directional distinction between the degrees of concreteness of the "familiar" term (the *vehicle*) of the metaphor and the one that is being characterized (the *tenor*).[2] This differential is what produces new knowledge as, according to this view, we acquire it in two manners: either through direct physical experience with our environment or through metaphorical understanding built upon some initial direct physical experience. For instance, we have the direct experience of the concept "up" and happiness is understood metaphorically through the metaphor "happy is up." This understanding is reflected in different expressions, such as "I'm feeling up today" or "I'm high as a kite," in which more abstract concepts are metaphorized and grasped in terms of less abstract ones. In Lakoff and Johnson's perspective (1980), metaphor is fundamentally directional: in the metaphorical pairing, one of the concepts is always better delineated and typically more concrete than the other.

> Human beings are constantly attempting to develop conceptions about the world, and as Cassirer (1946, 1955) and others have argued, they do so symbolically, attempting to make the world concrete by giving it form (Morgan 1980, 609–10).

2. METAPHORS AS INFLUENCERS

The asymmetric reduction of abstract to concrete, or at least "less abstract," is the key feature in the use of metaphors and models in philosophy and science, to reduce the unknown and the excessively complex to the known and relatively simple. Not all philosophers have seen this function as benign, though. For instance, empiricist philosophers mistrust metaphors and consider them confusing and misleading devices, which obfuscate the categorical distinctions between words. Metaphor is regarded as a matter of extraordinary language, a rhetorical flourish that, through emotion, misleads judgment and insinuate wrong ideas (Locke 1975, 508). Hobbes fears the "innumerable absurdities" that metaphors may produce (Hobbes 1962, Pt I, chap. 5). He considers the human conceptual system to be essentially literal and points out that "words proper" are adequate to express meaning, as a metaphor's meaning—when it has one—is its literal paraphrase.

The empiricist view, however, has been radically challenged by a series of novel approaches. The novelty of these approaches resides in regarding conceptual metaphors, ubiquitous in our everyday language, as the foundation of the entire cognitive system, construed as a web of interconnected prototypical metaphors. In the area of environmental sustainability, for instance, Larson points out the need of greater sensitivity to the presence and implications of metaphors, because we rely on them in our attempts to understand reality and because they are ubiquitous in the science we hear about every day (Larson 2011). Here we accept the view that, as metaphors are a matter of thought and not merely language, they play a role in shaping the way individuals approach and reason about complex issues. The use of certain metaphors in connection with a given topic creates a *conceptual domain*—a certain organization of human experiences; different metaphors generate different conceptual domains, which organize experiences, shape individuals' thoughts and language in different ways and, for instance, have been shown to influence differentially the individuals' framing of approaches to difficult social and policy problems.

This issue has notably been researched in relation to crime. In a famous experiment, Thibodeau and Boroditsky (2011) show how the metaphoric framing of crime as "a wild beast preying on the city" encouraged survey respondents to put forward policy solutions involving enforcement or punishment, and describing it as a "virus infecting the city" encouraged solutions involving social reforms. In the latter case, "virus infecting the city" identifies the source domain from which metaphorical expressions are drawn. The target domain is "crime," the concept—and the ideas connected to it—being characterized. The characterization of "crime" then emerges by conceptual mapping, through a systematic set of correspondences between constituent elements of the two domains. Using this analytical approach, new concepts are introduced in the target domain that did not exist before the linkage (Way 1994). In other words, elements of the domain of crime—the target domain—originate from the source domain, with concepts such as "prevention" and "cure" among them. This conceptual mapping created by the metaphor, which only partially is made explicit in the language, is then responsible for influencing the shaping of thoughts and attitudes in connection with the target domain.

3. METAPHORS AND MODELS IN SCIENCE

The specific role of metaphors and models in science has been the subject of lengthy discussions. These discussions, however, may be aligned to yet another polarization of views, namely, the polarization between the comparison and the interactive views (Montuschi 2000). The *comparison view*, typical of the logicist tradition, regards models as purely illustrative and possibly didactic, as well as epistemically void and dispensable. In this view, for example, the sentence "science changes by natural selection" should more correctly be rephrased as "scientific change is *like* change by natural selection," implying that a relation of resemblance holds between the mechanism we believe is at work in scientific change and the process known as "natural selection." The similarity is only intended for the clarification of a mechanism that is new, still unexplored, or more difficult to understand, thanks to a model that we assume is more familiar to the listener. The model, however, can be forgotten once the main subject has been well understood and formalized. This means that the limited scope and the ephemeral nature of the metaphor does not require a criticism of the adequacy of the analogies as full correspondence of features between the two domains, because the analogies are meant to be only illustrative. Thus, any evaluation can be only based on the grounds of correctness, for instance, on the right interpretation of the concepts in the two domains.

The alternative *interactive view* goes beyond the formalist and purely logical interpretation of scientific theories and incorporates the dynamism of language into the complexity of scientific change. As analogies are dynamically created by the metaphors used by the speakers to put two subjects side by side (Black 1962), it is the dynamics between the features of a familiar system with a well-consolidated and relatively unproblematic theory—the *explanans*—and those of an unfamiliar system, the *explanandum* that makes a model useful in developing or refining a theory (Hesse 1966). Between the two, a number of analogies and disanalogies will exist. The positive analogies will help to characterize the novel by its model, while it will be unclear or just unknown whether some properties of the model also belong to the *explanandum*. These undecided analogies constitute the set of neutral analogies that provide scientists with further material for research, and it is their potentially promising content that constitutes the fruitfulness of a model, that is its provision of research interest by suggesting how the theory may be extended (Achinstein 1968). In this view, models and metaphors do not merely unveil preexisting analogies but *create* analogies due to the interaction between the two subjects, with the neutral ones essential in stimulating scientific investigation.

4. KUHN

Thomas Kuhn maintained that "human cognition is governed at bottom by rhetorical relations of similarity, analogy, metaphor, and modeling rather than by logical relations and rules" (Nickles 2003, 6). He employed a number of metaphors and analogies to illustrate his account of scientific change and the concept of incommensurability, although he admitted they contributed to ambiguities and misunderstandings of his ideas (Hoyningen-Huene 1993, xv). Best known even to nonspecialists are probably those referring to a Gestalt switch and to a language change, while the analogy with organic evolutionary change is more specific and better developed. Kuhn highlighted several features of biological evolution as analogs to scientific change to both support the plausibility of some characteristics of scientific development he thought crucial for his new theory (Renzi 2009b; Renzi and Napolitano 2011) and, at a higher level, illustrate the common features generated by these analogues in two instances of theory change: the transitions between pre- and post-Darwin evolution and the transition between pre- and post-Kuhn scientific change (Reydon and Hoyningen-Huene 2010). For instance, he used the evolutionary analogy to show how his whole concept of scientific change is, ultimately, not a relativistic one:

> Scientific development is, like biological, a unidirectional and irreversible process. Later scientific theories are better than earlier ones for solving puzzles in the often quite different environments to which they are applied. That is not a relativist's position, and it displays the sense in which I am a convinced believer in scientific progress (Kuhn 1970, 206).

This continuity, implied by the possibility to compare theories in their problem-solving power, fills the gap of incommensurability to the extent of showing the progressive nature of scientific development. This progression, however, is not directed toward anything but manifests itself in the form of improvement from the status quo. This corresponds, in Kuhn's evolutionary analogy, to the progressive but nondirected nature of biological evolution, and the two processes, via corresponding selective mechanism, produce analogous results. Both natural selection and "revolutionary selection" let the most viable of all possible alternatives, the one that is most "fit" for a particular, historical situation, survive.

> The resolution of revolutions is the selection by conflict within the scientific community of the fittest way to practice future science. The net result of a sequence of such revolutionary selections, separated by periods of normal research, is the wonderfully adapted set of instruments we call modern scientific knowledge (Kuhn 1970, 172).

In this view, newly formulated theories that compete with the established, paradigmatic ones correspond to the emerging mutations occurring in competing organisms.[3]

It is worth mentioning that many others employed evolutionary analogies to describe or explain scientific change in even stronger terms.[4] Karl Popper persistently made the strong claim that science *evolves by* natural selection (although not always consistently):

> The evolution of scientific knowledge is, in the main, the evolution of better theories. This is, again, a Darwinian process. The theories become better adapted through natural selection: they give us better and better information about reality (they get nearer and nearer to the truth). All organisms are problem solvers: problems arise together with life (Popper 1984, 239).

David Hull also argues for natural selection and "scientific selection," as well as for the selective mechanisms acting in society and more generally in conceptual evolution, to be regarded as instantiations of the same process (Hull 1982, 2006). Kuhn, instead, preferred to focus on individual evolutionary concepts, show the similarities with the analogous concepts he saw in the development of science and, in doing so, support some of the controversial aspects of his theory. For instance, it seems clear that one crucial result of Kuhn's use of an evolutionary analogy is the presentation of the possibility of progressive but nondirected selective processes, as well as the solution to the issue of relativism otherwise implied by his incommensurability thesis.

5. SCIENTIFIC CHANGE BY ORTHOGENESIS?

While evolutionary change has been the modality preferred by many to characterize or illustrate metaphorically their theories of scientific change, it is probably not the accepted version of organismic evolution that provides the best fit: the orthogenetic hypothesis may be a better analogy than Darwinism insofar as describing the process of scientific change is concerned. The orthogenetic hypothesis of biological evolution is the basis of a number of theories embracing a nonadaptive idea of directed evolution. In the nineteenth century, these theories tried to undermine the tenets of Darwinism and the role of natural selection in organismic evolution, by incorporating a notion of progress. This small cluster of theories was never fully developed and was eventually made obsolete when the combination of natural selection, Mendelian genetics, and a better reading of the fossil records succeeded in explaining several difficulties of the original Darwinian exposition (Futuyma 1998). While Wilhelm Haacke[5] postulated a progressive tendency in nature

toward higher forms of complexity, possibly with temporary reversions in the
trend, it was Theodor Eimer who popularized the term "orthogenesis" at the
end of the nineteenth century (Bowler 1979, 40). He maintained that variations are not random but directed by forces regulated and ultimately directed
by the internal constitution of the organism, which responds to environmental
stimuli. Similar views were held by Thomas Henry Huxley (1978): although
the environment can influence modifications in a species, it is the internal
constitution of the species that leads the pattern of its development and the
resulting "inclination" can be considered its main directing element.

One of the issues regarding Darwinian selective models of scientific
change concerns the emergence of new ideas in the scientific community,
which are analogous to the occurrence of biological variations. In natural
selective models, there is no correlation between occurring variations and
the pressures of the organism's environment or its needs. On the other hand,
changes in scientific ideas, which emerge as scientists try to solve problems,
are the results of arguments and debates, the answers to the specific needs
of a scientist or group of scientists who have been seeking a solution to a
problem. This process of scientific change is more similar to orthogenesis
than to Darwinian evolution.[6] As orthogenetic variations are regulated by
constitutive factors in the organisms, which establish a developmental direction, similarly in science the existing knowledge regulates and directs the
pattern of development. The difficulties around the progressive nature of the
scientific enterprise, which stem from the attempt to account for both scientific and biological evolution in terms of the same mechanism of Darwinian
selection, dissolve in the case of orthogenesis, as the overall trend of orthogenetic evolution is progressive, because of the production of progressively
more complex structures in place of simpler ones. On the other end, while
scientific knowledge changes new directions of development become available, and this also occurs in orthogenesis:

> In my view development can take place in only few directions because the
> constitution, the material composition of the body, necessarily determines such
> directions and prevents indiscriminate modification.
>   But through the agency of outward influence the constitution must gradually
> get changed. The organisms will thus acquire more and more in a manner har-
> monising with their specific individuality—and so new directions of develop-
> ment will be produced.[7]

In other words, an analogous mechanism to that proposed by Eimer may
occur in science because, although variation is directed, new directions of
scientific development can be produced. The orthogenetic hypothesis may
even be more adequate in accommodating further characteristics advanced

by other philosophers for the mechanism of scientific change, suggesting that this hypothesis may be a better generalization of the features possibly in place. For instance, Imre Lakatos' approach may be considered as an attempt to salvage the reasonable aspects of the theories proposed by Popper and Kuhn. His central argument (1970) of the research program as the basic unit of scientific evaluation, as opposed to single theories or ideas, and his distinction between the core ideas characterizing a research program and its protective belt of auxiliary hypotheses, which shield the hard core from falsification, have a dual role: to soften Popper's concept of falsifiability and to restore some of the rationality lost in Kuhn's view of theory change and incommensurability. For the evolutionary analogy, Lakatos' account of the heuristic embedded in the core of a research program has special importance. The heuristic is what provides direction to the modifications in the protective belt, which are instigated by the empirical developments, and suggests to scientists what the paths of research worth pursuing are and which routes they should avoid.

Like the organisms' constitution for evolution proposed by the orthogenetic hypothesis, the heuristic of a research program is the most stable developmental drive for scientific change. Further, organic evolution by orthogenesis is the result of the interaction between the embodied laws of development and the contingent external environment and, similarly, the heuristic and the core ideas do not direct the development of the research program alone. This development is also informed by the results of tests and experiments performed on nature and, under the influence of external conditions, both the biological constitution and the scientific hard core may (eventually) change. When this occurs, a new species emerges or, in the analogy, a new research program is generated.[8]

6. HOW METAPHORS SHOULD BE USED

Although Kuhn's rendering of scientific change has been analyzed and criticized in depth and from different perspectives, especially regarding progress and incommensurability, here we have proposed that, in the light of the current approach to metaphors, more insights may be provided by analyzing the metaphors used to support or illustrate it, in the framework of the general scientific change process. We have mentioned how, in recent research, it has been shown the metaphorical framing is ubiquitous in human communication and that, far from illustrative and dispensable devices, metaphors may play an essential role in the cognitive processes of comprehension and ideation. While metaphors and figurative models are, ultimately, noninvalidable in the sense that they do not provide a formal description of a given portion of reality susceptible to testing, they do provide scientists and philosophers with a direction in their production of ideas and theories that may turn out valuable or not. Regardless of how a metaphorical framing may emerge in the mind of a researcher, once it is proposed that connections emerge autonomously and may direct how the problem is approached. Sometimes the metaphorical framing is particularly close to the discipline of the proponent and it dominates with particular strength. For instance, we have mentioned how David Hull has been particularly meticulous in developing a selective model that included both science and organismic change. Hull, a taxonomist, analyzed in great detail the long-standing diatribes between pheneticists and cladists to illustrate the changes occurring within the science of taxonomy. Because of this "choice" and because taxonomist classifications are conceived more as the construction of the taxonomist's view of the world, rather than an attempt to reflect reality (Ruse 1989), Hull's account of scientific methodology is reduced to a sociological phenomenon that negates any special status to science and scientific products.

This phenomenon, that is, the projection of undesired (or undesirable) features onto an analyzed system caused by the metaphorical framing or model of choice, can be found in other domains as well. Gareth Morgan found that

> schools of thought in organization theory are based upon the insights associated with different metaphors for the study of organizations, and [. . .] the logic of metaphor has important implications for the process of theory construction. The use of a metaphor serves to generate an image for studying a subject. This image can provide the basis for detailed scientific research based upon attempts to discover the extent to which features of the metaphor are found in the subject of inquiry (Morgan 1980, 611).

Morgan analyzed several metaphors employed in organization theory including "organism," "machine," "cybernetic system," "theatre," and "ecological system," among others. Morgan observes that the features of a chosen metaphor provide

> the focus of detailed theory and research, often to the exclusion of all else. Such a perspective results in a premature closure in both thought and inquiry. Schools of theorists committed to particular approaches and concepts often view alternative perspectives as misguided, or as presenting threats to the nature of their basic endeavor (Morgan 1980, 613).

In other words, while metaphors can be valuable in providing hints and directions toward promising routes of research and exploration, they may also be dangerous with the filtering effect that the directional of their very nature embeds. At the time, Morgan found that two predominant metaphors had shaped the mainstream view in organization theory: "the organization is a machine" and "the organization is an organism." He notes that "whereas in the machine metaphor the concept of organization is as a closed and somewhat static structure, in the organismic metaphor the concept of organization is as a living entity in constant flux and change, interacting with its environment in an attempt to satisfy its needs" (Morgan 1980, 614). Further, in Morgan's view, the consequence is that theorists "fail to understand that the apparent order in social life is not so much the result of an adaptive process or a free act of social construction, as the consequence of a process of social domination" (Morgan 1980, 619–20). This is an interesting point. As an evolutionary analogy may also emphasize the adaptive aspects of change in the scientific enterprise and downplay the effects of social domination, power (and funding), and control, it is evident that the choice of metaphors to illustrate scientific change may have unintended, implicit consequences.

In conclusion, a good approach to the use of metaphors in philosophy of science in order to analyze scientific change should be a cautious one. If we think of a good analogy as "simply one that contributes to the solution of a given set of problems" (Thagard 1993, 102), the best approach is possibly to use as many different ones as possible to fit separate aspects and domains of the process. After this "metaphorical map" has reached a good coverage of the whole domain, it is possible that the various metaphors and analogies have converged independently on a single "source" domain, thereby suggesting that the same fundamental mechanisms may be operating, at least partially, in both domains. At that point, a more formal, or at least intersubjective, methodology should be employed to evaluate the possibility. For instance, a type hierarchy approach may be used (Renzi 2009a; Renzi and Napolitano 2011) to show that organismic evolution and scientific change cannot be regarded as two instances of the same general process, with the abandoned[9] orthogenetic hypothesis being a better candidate than the current theory of evolution. At this stage of the analysis, it is further scrutiny of the points of disagreement and disanalogies, such as intentionality and direction, and not of those left open, that will provide, possibly by sourcing new metaphors and analogies from different domains, new fruitful avenues of exploration.

7. CONCLUSION

In this chapter, we have outlined the cognitive view of metaphors and mentioned how metaphors and analogies are increasingly being regarded as being at the basis of cognition, involved not only in the process of understanding and simplification but also in directing or at least influencing ideation, attitude, and possibly behavior. In the case of Kuhn's and other philosophers' evolutionary accounts of scientific change, this implies that special attention should be paid not only to the epistemological framework of their models but also to the perspectives that these accounts may force onto those who also follow and, possibly, further develop them. As examples of possible perspectives, we have mentioned two cases: the overemphasis of adaptive mechanisms, which downplays the role that social domination exerts in the scientific enterprise; and the complete reduction of science to a sociological phenomenon, which renders scientific products devoid of any special status. In conclusion, as metaphors may exert considerable power in shaping the direction that the theories they are supposed to illustrate may or may not take, philosophers should be fully aware of how metaphors are employed in their explanations and avoid the risk of being blinded by their use.

NOTES

1 In a simile, the similarities involved in the comparison are clearly defined, and terms such as "like," "as," and "not unlike" are present in the statement of the comparison.

2 The word "tenor" and the term "vehicle" were introduced by Richards (1936). Tenor is the underlying idea or the principal subject, while the vehicle is what is attributed metaphorically to the tenor. For instance, consider the case "men are wolves"; "men" is the tenor and "wolves" the vehicle.

3 See also Marcum (2015) for a defense of evolutionary models of scientific change, as well as for an account of Kuhn's trajectory from a historical to an evolutionary philosophy of science.

4 See Renzi and Napolitano (2011) for a longer analysis.

5 Haacke, W. (1893) *Gestaltung and Vererbung*, O. W. Nachforger, Leipzig quoted in Grehan and Ainsworth (1985).

6 See Renzi and Napolitano (2011) for a more formal argument based on type hierarchies.

7 Eimer G. H. T. (1898) *On orthogenesis and the Impotence of Natural Selection in Species Formation*, trans. J. M. McCormack, Chicago: Open Court, 22 quoted in Bowler (1979, 50–51).

8 It is worth noting that the level of selection (ideas, theories, or research programs) is not an issue here. In organismic evolution, some classify as "natural selection" only the selective process acting at the level of genes, genotypes, and individual organisms, whereas others also include the level of groups, such as species and populations (see Futuyma 1998 for more details). This, however, does not affect what are regarded as the characteristic features of natural selection.

9 More sophisticated (and diluted) orthogenetic views are still supported by some researchers (Grehan and Ainsworth 1985), although they do not appear to have many followers.

REFERENCES

Achinstein, Peter. 1968. *Concepts of Science.* Baltimore, MD: John Hopkins Press.

Aristotle. 1941. "Poetics." In *The Basic Works of Aristotle,* edited by Richard McKeon, 1455. New York: Random House.

Black, Max. 1962. *Models and Metaphors.* Ithaca, NY: Cornell University Press.

Bowler, Peter, J. 1979. "Theodor Eimer and Orthogenesis: Evolution by 'Definitely Directed Variation.'" *Journal of the History of Medicine and Allied Sciences* 34, 40–73.

Cassirer, Ernst. 1946. *Language and Myth.* New York: Dover.

Cassirer, Ernst. 1955. *The Philosophy of Symbolic Forms, vols. 1–3.* New Haven: Yale University Press.

Davidson, Donald. 1978. "What Metaphors Mean." *Critical Inquiry* 5, 31–47.

Futuyma, Douglas, J. 1998. *Evolutionary Biology.* Sunderland, MA: Sinauer Associates.

Grehan, John R., and Ainsworth, Ruth. 1985. "Orthogenesis and Evolution." *Systematic Zoology* 34(2): 174–92.

Hesse, Mary, B. 1966. *Models and Analogies in Science.* Notre Dame, IN: University of Notre Dame Press.

Hobbes, Thomas. 1962. "The Leviathan." In *The English Works of Thomas Hobbes of Malmesbury*, edited by Sir William Molesworth, Vol. III. Aalen, Germany: Scientia.

Hoyningen-Huene, Paul. 1993. *Reconstructing Scientific Revolutions: Thomas S. Kuhn's Philosophy of Science.* Chicago: University of Chicago Press.

Hull, David, L. 1982. "The Naked Meme." In *Learning, Development, and Culture*, edited by H. C. Plotkin, 273–327. New York, NY: Wiley.

Hull, David L. 2006. "The Essence of Scientific Theories." *Biological Theory* 1, 17–19.

Huxley, Thomas, H. 1978. "Evolution in Biology." Reprinted in (1983) *Collected Essays II, Darwiniana*, 187–226. London: Macmillan.

Johnson, Mark, ed. 1981. *Philosophical Perspectives on Metaphor.* Minneapolis: University of Minnesota Press.

Kuhn, Thomas, S. 1970. *The Structure of Scientific Revolutions*, Second Edition, Enlarged. Chicago: University of Chicago Press.

Lakatos, Imre. 1970. "Falsification and the Methodology of Scientific Research Programmes." In *Criticism and the Growth of Knowledge*, edited by I. Lakatos and A. E. Musgrave, 91–195. Cambridge: Cambridge University Press.

Lakoff, George. 1993. "Contemporary Theory of Metaphor." In *Metaphor and Thought*, edited by Andrew Ortony, 202–51. Cambridge University Press.

Lakoff, George, and Johnson, Mark. 1980. *Metaphors We Live by.* Chicago: University of Chicago Press.

Larson, Brendon. 2011. *Metaphors for Environmental Sustainability: Redefining Our Relationship with Nature.* New Haven, CT: Yale University Press.

Locke, John. 1975. *An Essay Concerning Human Understanding*, edited by P. H. Nidditch. Oxford: Clarendon Press.

Marcum, James A. 2015. "What's the Support for Kuhn's Incommensurability Thesis? A Response to Mizrahi and Patton." *Social Epistemology Review and Reply Collective* 4, 51–62.

Martinich, A. P. 1984. "A Theory for Metaphor." *Journal of Literary Semantics* 13, 35–56.

Montuschi, Eleonora. 2000. "Metaphor in Science." In *A Companion to the Philosophy of Science*, edited by W. H. Newton-Smith, 277–282. Oxford: Blackwell.

Morgan, Gareth. 1980. "Paradigms, Metaphors, and Puzzle Solving in Organization Theory." *Administrative Science Quarterly* 25, 605–22.

Nickles, Thomas, ed. 2003. *Thomas Kuhn*. Cambridge: Cambridge University Press.

Popper, Karl, R. 1974. *Conjectures and Refutations: The Growth of Scientific Knowledge*. London: Routledge and Kegan Paul.

Popper, Karl, R. 1984. "Evolutionary Epistemology." In *Evolutionary Theory: Paths into the Future*, edited by J. W. Pollard, 239–255. Chichester, England: John Wiley &Sons Ltd.

Renzi, Barbara, G. 2009a. "A Type Hierarchy for Selection Processes for the Evaluation of Evolutionary Analogies." *Journal for General Philosophy of Science* 40, 311–36.

Renzi, Barbara, G. 2009b. "Kuhn's Evolutionary Epistemology and Its Being Undermined by Inadequate Biological Concepts." *Philosophy of Science* 76, 143–59.

Renzi, Barbara, G., and Napolitano, Giulio. 2011. *Evolutionary Analogies: Is the Process of Scientific Change Analogous to the Organic Change?* Newcastle: Cambridge Scholars Publishing.

Reydon, Thomas, A. C., and Hoyningen-Huene, Paul. 2010. "Discussion: Kuhn's Evolutionary Analogy in *The Structure of Scientific Revolutions* and *The Road since Structure*." *Philosophy of Science* 77, 468–76.

Richards, I. A. 1936. *The Philosophy of Rhetoric*. Oxford: Oxford University Press.

Ricoeur, Paul. 1977. *The Rule of Metaphor*. Toronto: University of Toronto Press.

Rumelhart, David, E. 1993. "Some Problems with the Notion of Literal Meanings." In *Metaphor and Thought*, edited by Andrew Ortony, 71–82. Cambridge University Press.

Ruse, Michael. 1989. "Great Expectations." *The Quarterly Review of Biology* 64, 463–68.

Thagard, Paul. 1993. *Computational Philosophy of Science*. Cambridge, MA: The MIT Press.

Thibodeau, Paul, H., and Boroditsky, Lera. 2011. "Metaphors We Think with: The Role of Metaphor in Reasoning." *PLOS ONE* 6(2): e16782.

Way, Eileen, C. 1994. *Knowledge Representation and Metaphor*. Oxford: Intellect Books.

Chapter 10

Beyond Kuhn

Methodological Contextualism and Partial Paradigms

Darrell P. Rowbottom

1. KUHN'S IMAGE OF SCIENCE: NORMAL SCIENCE, CRISIS, AND EXTRAORDINARY SCIENCE

Kuhn's view of science is as follows.[1] Science involves two key phases: normal and extraordinary. In normal science, *disciplinary matrices* (DMs) are large and pervasive.[2] DMs involve "beliefs, values, techniques, and so on shared by the members of a given community" (Kuhn 1996, 175). "And so on" is regrettably vague, but Kuhn (1977; 1996) mentions three other key elements: symbolic generalizations (such as $F=dp/dt$), models (such as Bohr's atomic model), and exemplars (which I explain below). These components of DMs overlap somewhat. For instance, symbolic generalizations may feature in techniques and beliefs, and models may exhibit values.

To be a (genuine) scientist, in the normal science phase, is to *puzzle solve* within the boundaries of the DM. It is to buy into the ruling dogma (Kuhn 1963) and to accept that "failure to achieve a solution discredits only the scientist . . . 'It is a poor carpenter who blames his tools'" (Kuhn 1996, 80). Puzzle solving involves a wide variety of activities, including bringing observations into closer agreement with theories (e.g., by altering auxiliary assumptions), articulating existing theories (e.g., by measuring constants), and classifying things, or kinds of thing, in line with the DM (e.g., measuring the magnitude of a star or the Young's modulus of an alloy).[3] It can sometimes resemble exploration, but isn't genuinely exploratory.

Kuhn (1996, 187) declares that exemplars are "the central element of . . . the most novel and least understood aspect" of normal science. These are "concrete puzzle-solutions which, employed as models or examples, can replace explicit rules" (Kuhn 1996, 175) for puzzle solving. They typically involve shared theories and models—perhaps also shared symbolic generalizations—and exhibit shared values. They invariably use shared techniques. In essence, exemplars provide templates for tackling new puzzles, as well as means by which to assess potential solutions to puzzles. Moreover, they can help in the identification of some new puzzles. Using them effectively involves spotting similarities, which is more practicable than attempting to internalize and follow rules.[4] Imagine, for instance, that a student has worked through a problem concerning circular satellite motion around Earth (involving classical mechanics and Newton's law of gravitation). She is subsequently taught Coulomb's law of electrostatic attraction. Provided she spots the relevant similarities—for example, the shared form of Newton's law and Coulomb's law—then she will be well equipped to consider how a charged massless body might circularly orbit an oppositely charged massless body. It might also naturally occur to her that she could potentially deal with more complex puzzles, involving stable circular orbits with both gravitational and electrostatic forces present, in a similar way.

Kuhn holds that the practice of normal science will almost inevitably lead to the appreciation of various anomalies or "violations of [DM-based] expectations" (Kuhn 1996, ix). Especially noteworthy are puzzles that are taken to be significant but unsolved. These can be tolerated, to the extent that they might reasonably be expected to be transient:

> Failure with a new sort of problem is often disappointing but never surprising. Neither problems nor puzzles yield often to the first attack . . . There are always some discrepancies. Even the most stubborn ones usually respond at last to normal practice (Kuhn 1996, 75, 81).

However, long-term anomalies may eventually begin to shake the confidence of scientists in their DM: "insecurity is generated by the persistent failures of the puzzles of normal science to come out as they should" (Kuhn 1996, 68). At some point—whether this happens will depend on a variety of factors, some of which might be external to science—a "crisis" may result: "when confronted by . . . severe and prolonged anomalies . . . [scientists] may begin to lose faith and then to consider alternatives, [although] they do not renounce the paradigm [DM] that has led them into crisis" (Kuhn 1996, 77). They do not renounce the DM because there is no live alternative, and because a DM is required:

> Once a first paradigm through which to view nature has been found, there is no such thing as research in the absence of any paradigm. To reject one paradigm without simultaneously substituting another is to reject science itself. That act reflects not on the paradigm but on the man. Inevitably he will be seen by his colleagues as "the carpenter who blames his tools" (Kuhn 1996, 79).

However, crisis leads to *extraordinary* science in so far as it directs efforts toward removing the anomaly or anomalies at its heart—in so far as it focuses scientists, especially eminent scientists, on solving the key unsolved puzzle or puzzles, which come to be perceived as *problematic* in character: "more and more attention is devoted . . . by more and more of the field's most eminent . . . If it still continues to resist, as it usually does not, many of them may come to view its resolution as *the* subject matter of their discipline" (Kuhn 1996, 82–83).

Initially, Kuhn says, attempts to solve such problems will proceed from within the DM. But over time, as attempts fail, more and more liberties will be taken, and different articulations or versions of the DM will arise. And hence the (implicit) rules for puzzle solving will be relaxed. Kuhn (1996, 83) puts it as follows:

> Through this proliferation of divergent articulations (more and more frequently they will come to be described as *ad hoc* adjustments), the rules of normal science become increasingly blurred. Though there still is a paradigm [DM], few practitioners prove to be entirely agreed about what it is.

Yet extraordinary science may go even further in so far as it may result in "the willingness to try anything . . . [and] the recourse to philosophy and to debate over fundamentals" (Kuhn 1996, 91). Thus, it is implicit in Kuhn's writing—on a note to which we will return—that extraordinary science, like crisis, is a matter of degree.

The presence of crisis and the shift to extraordinary science sets the stage for a new DM to emerge. But this is not the only possible outcome. Sometimes normal science under the existing DM saves the day. On other occasions, no new candidate DM is found in a reasonably timely fashion, and work in the area is suspended. Suspension might be temporary—new technology might pave the way for progress, for instance—or permanent.

1.1. From the Descriptive to the Normative

Thus far, we have seen how Kuhn *describes* science, based largely on his historical studies. (It should be remembered, though, that observation in the history of science is plausibly as theory laden as observation in science.) Kuhn doesn't stop there, however, because he thinks that one can derive some "oughts" from the "ises." In his own words, "the descriptive and the normative are inextricably mixed" (Kuhn 1970b, 233).[5] He also explicitly states, in the postscript to *The Structure of Scientific Revolutions*, that he does "present a viewpoint or theory about the nature of science . . . [which] like other philosophies of science . . . has consequences for the way in which scientists should behave if their enterprise is to succeed" (Kuhn 1996, 207). His justification for this move appears to be evolutionary in character; he claims—but does not argue—that the methods used by scientists "have been developed and selected for their success" (Kuhn 1996, 208). And on this basis, he thinks his "descriptive generalizations are evidence for the theory precisely because they can also be derived from it" (Kuhn 1996, 208). Kuhn (1963) makes a similar move in his argument for the value of dogma and indoctrination (of a fashion)—including intentional distortion and misrepresentation of the history of science—in science education. In short, he claims that science education *as he describes it* is well suited to the task of creating good ("normal") scientists.[6]

But why does Kuhn think that science *as he describes it* is good science? The key to the answer lies, as I've already intimated, in the normal phase, which Kuhn (1970a, 6) takes to be more central, in characterizing science, than the extraordinary phase:

> it is for the normal, not the extraordinary practice of science that professionals are trained . . . . If a demarcation criterion exists (we must not, I think, seek a sharp or decisive one), it may lie in just that part of science.

Normal science is crucial, on Kuhn's view, for several *strategic* reasons.[7] The following three are key, and involve not only its immediate products but also its role in fomenting crises and thus extraordinary science (of a particular kind). First, the confidence (or even faith) involved means that the group is focused on completing similar tasks, including tasks with practical benefits, and is able to concentrate its efforts on completing those tasks. In the absence of a DM to which the community is *strongly* committed, hypercriticism—involving prolonged and unproductive squabbles about fundamental matters (of a metaphysical variety, *inter alia*)—is a serious potential problem. Kuhn puts it as follows. On the one hand, "Because they can ordinarily take current theory for granted, exploiting rather than criticizing it, the practitioners of mature sciences are freed to explore nature to an esoteric depth and detail otherwise unimaginable" (Kuhn 1970b, 247). And, on the other hand, "The scientist who pauses to examine every anomaly he notes will seldom get significant work done" (Kuhn 1996, 82).[8]

Second, on a related but subtly different note, it is far more efficient to solve puzzles using existing techniques, assuming they are fit for the purpose, than it is to endeavor to solve them in another fashion. In the words of Kuhn (1996, 76):

> So long as the tools a paradigm [DM] supplies continue to prove capable of solving the problems it defines, science moves fastest and penetrates most deeply through confident employment of those tools. The reason is clear. As in

manufacture so in science—retooling is an extravagance to be reserved for the occasion that demands it.

Third, and finally, Kuhn believes that long periods of normal science are useful in so far as they are liable to indicate the true limits of the ruling DM, and especially to identify puzzles that are genuinely insoluble within the constraints of said DM. Kuhn (1996, 65) puts it so: "By ensuring that the paradigm [DM] will not be too easily surrendered, resistance guarantees that scientists will not be lightly distracted and that the anomalies that lead to paradigm [DM] change will penetrate existing knowledge to the core." Elsewhere, he amplifies the point as follows: "Because . . . exploration [of the DM] will ultimately isolate severe trouble spots, they [normal scientists] can be confident that the pursuit of normal science will inform them when and where they can most usefully become Popperian critics" (Kuhn 1970b, 247). In short, that's to say, Kuhn's idea is that extraordinary science *coming after an extended period of normal science* will tend to be superior to the similar kind of approach that may occur in nascent (or proto-) science, because its focus will be sharper and more appropriate. This point is especially persuasive if one accepts that many practical problems on which science bears—problems of engineering, for instance—may persist through DM-change.

This concludes my account of Kuhn's (early-middle period) image of science. To the extent that it is vague or silent in several significant respects—for example, on how different DMs should be distinguished from variants of the same DM, on how many people it takes to form (and thus what legitimately counts as) a community, on when and how precisely normal science terminates and extraordinary science begins, and on what socioeconomic conditions are necessary (or at least sufficient) for science to occur—I typically hold Kuhn, rather than myself, responsible.

I will cover some such defects of Kuhn's image in the next section. I shall also cover several problems with the most developed aspects of his account. In the third section, I shall propose a means by which to address these problems. I will explain how this involves rejecting the Kuhnian image while retaining some elements thereof.

2. ASSESSING KUHN'S IMAGE

Kuhn's image is problematic in both the descriptive and normative dimensions. Concerning the former, one may legitimately doubt—as Toulmin (1970) does—whether all science or even most science fits the Kuhnian mould. Consider a domain in a mature science, in a particular year, picked at random: atomic theory in 1910, biomechanics in 1970, and so forth.

Should we expect this to exhibit normal science (of a crisis-free variety)? More trenchantly—since the former expectation would rest on a probability claim—if we make many such random picks will we find that the relative frequency of normal science (of a crisis-free variety) is considerably greater than one half, as Kuhn suggests? The simple truth is that we don't know, as we didn't when Toulmin (1970) was writing. Moreover, even if this research were to be assiduously conducted, legitimate disagreements would arise about which picks instantiated normal science. Recall, for example, Kuhn's (1996, 79, 83) claims that there is "no such thing as research in the absence of any paradigm [DM]" and that there may be a "paradigm [DM] . . . [although] few practitioners prove to be entirely agreed about what it is." Combine these with his insistence that there are *versions* of theories and DMs—which features in claims such as: "by proliferating versions of the paradigm [DM], crisis loosens the rules of normal puzzle-solving" (Kuhn 1996, 80)—and the difficulties should be evident. One can't look to what the practitioners think to determine whether there's a single disciplinary matrix (although one *might* look to what they do, in so far as this is historically possible). Moreover, one can't be sure that differences in approach—even apparently pronounced differences—entail differences in disciplinary matrix, because Kuhn gives no guidance about how to judge when two different collections of shared commitments—to exemplars, methods, values, and so forth—constitute versions of the same disciplinary matrix, as opposed to different disciplinary matrices.

The likely riposte to this is as follows: it's a straightforward matter of similarity. However, this will not do. Even granting that there is a way to select a principled unique measure of similarity—and this is far from obvious (in so far, e.g., as there may be no context-independent fact of the matter about the extent to which overall resemblance obtains between any two things)[9]—it is evident that *similarity comes in degrees*.[10] Thus, it is implausible that there is a "critical point" at which a single change to a disciplinary matrix makes it cease to be a *version* of its predecessor, rather than an entirely different disciplinary matrix. Indeed, the very notion that disciplinary matrices may be *versions* of one another appears to be arbitrary. Why think this rather than simply that they may be similar to greater or lesser degrees? The problem for Kuhn is that the retreat to discussing only degrees of similarity allows that many changes might legitimately be made, and the very division between normal and extraordinary science is under threat. In its place is a continuum. It is natural to think, moreover, that this continuum will—and to foreshadow the discussion given later, should—vary according to the scientists' degree of confidence in a DM, or confidence in a group of available DMs, being able to solve the set of available or pressing puzzles (some of which might count as anomalies).

Two further remarks are in order before I proceed to discuss the normative aspects of Kuhn's account, which will be my focus in the remainder of this chapter. First, Kuhn emphasizes degrees of change only when doing so tends to suit his narrative about science, rather than the narratives of his opponents (e.g., Popper). For example, he writes: "the existence of a crisis does not by itself transform a puzzle into a counterinstance. There is no such sharp dividing line" (Kuhn 1996, 80). That's to say, he emphasizes the role of degrees to which anomalies are seen as significant (which one might take to bear on degrees of confidence in any DM). Second, note that many of Kuhn's claims about DMs are too strong, because they neglect the possibility of *sets* of active DMs. Consider, for example, Kuhn's (1996, 65) claim that "anomaly appears only against the background provided by the paradigm [DM]." Nowhere does Kuhn argue that anomaly cannot appear instead against the background provided by a group of active DMs. And it might, in so far as each of a group of active DMs—in one and the same community—may involve some of the same puzzles. One could *declare* any such DMs to be "versions of the same DM," but we have already seen that this is arbitrary.

Now let's consider Kuhn's image of science from a normative perspective, even granting that it is descriptively accurate. One implausible aspect lies precisely in the putative link between the descriptive and the normative. Wouldn't it be a miracle if mature science *as it is done* were optimal, or at least approximately so, from a "strategic" perspective? Appeal to an evolutionary analogy—recall Kuhn's (1996, 208) suggestion that the methods used by scientists "have been developed and selected for their success"—will not do the trick. Not all mutations have been explored, and the possible mutations can change over time; we now have access to technologies we previously didn't, for example. Moreover, what's perceived as success can change over time—curiously, Kuhn argues for this—and thus what was previously selected against might not be selected against were it to later reappear (and it might be unlikely to so reappear, if it has been forgotten). Analogously, we know that we are far from optimal in many respects despite having gone through (along with our evolutionary ancestors) many more generations than science (or its methods, theories or matrices). One of the most striking recent illustrations of this was the invention of an exoskeleton that "consumes no chemical or electrical energy and delivers no net positive mechanical work, yet reduces the metabolic cost of walking by $7.2 \pm 2.6\%$ for healthy human users under natural conditions" (Collins et al. 2015, 212).

Perhaps Kuhn would respond to this by granting that improvements are possible, while denying that any of these would involve changing the process he sees as *central* to, and *necessary for*, good science—the cycle of normal science (i.e., dogmatic puzzle solving under a single DM), crisis, extraordinary science, and revolution. Yet how does Kuhn purport to know this, even assuming that he has spotted a genuine pattern in the history of science? There are numerous alternative processes, many of which, no doubt, Kuhn didn't conceive of. And this throws serious doubt on his remarkably strong claims about the impossibility of achieving many things in any other way, which appear throughout *The Structure of Scientific Revolutions* (and which I've already criticized to some extent, with reference to counterfactual history, in endnote 8). Here's a selection:

> during the period when the paradigm [DM] is successful, the profession will have solved problems that its members . . . would never have undertaken without commitment to the paradigm [DM] (Kuhn 1996, 24–25).
>
> within those areas to which the paradigm [DM] directs the attention of the group, normal science leads to a detail of information and to a precision of the observation-theory match that could be achieved in no other way (Kuhn 1996, 65).
>
> [there are puzzles] for whose very existence the validity of the paradigm [DM] must be assumed (Kuhn 1996, 80).

I have already said something about the final claim—the quick refutation, recall, is that one can recognize and even solve puzzles in frameworks one doesn't think are valid, such as deviant logics—so will now focus on the previous two. Imagination suffices to see their probable falsity. Consider the first. Imagine a community of scientists who think that the DM they use is invalid (whatever exactly that means). They have no dogmatic commitment to it. Perhaps they think it's dumb. But they're prisoners. Slaves. Their captors insist that the scientists work on this DM, or face terrible punishments. The captors need not even be committed to the DM. They could have many different groups of scientists held captive, each working on a different DM, in order to hedge their bets. They could be after a particular practical outcome, such as the development of a powerful new weapon. They could also be knowledgeable enough to do the work themselves—this would explain their ability to design the DMs—but prefer only to oversee it, given their wicked ways.

Now consider the second claim. The example discussed previously has already also refuted this, provided normal science fundamentally involves dogmatic commitment, as it does on Kuhn's view. But even if we relax this constraint, other plausible counterexamples are easy to come by. Consider, for example, a small group of dissident scientists, who refuse to use the dominant DM in their area despite the absence of crisis. Instead, they work together to develop and use alternative approaches. (If preferred, one can think that they have their own DM. But Kuhn *doesn't* advocate two DMs existing in the same area simultaneously—or even DM variants existing in the same area simultaneously—in the absence of crisis.) Their twin aims are to show that the dominant DM should be replaced and to solve some pressing practical problems that its users have not yet been able, although they expect eventually to be able, to solve. Why should we think that this dissident group will not acquire the same "detail of information" or "precision of observation-theory match" as the orthodox community? Why might they not strive for and achieve more, per capita per unit time, than the orthodox community? They might be especially driven by the fact that their results may have to be *more* precise than those of the orthodox group, in order for their work to be taken seriously (by third parties, perhaps, in the first instance).

The only reply that appears feasible—and then, only prima facie—is that if the dissident scientists joined the orthodox community, that community would do better (in terms of "detail of information" acquired and "precision of observation-theory match").[11] I suppose this may seem plausible from a highly simplistic perspective, according to which progress made is directly proportional to time spent. But this perspective is fundamentally flawed. Productivity per capita per unit time depends on numerous factors, including group cohesion and structure, the roles of and capacities of group members, and so on. And in the example discussed previously, attempts by the dissident scientists to integrate might be disastrous. They might prove to be somewhat disruptive influences in several respects, even if they did their best to toe the line. They might hamper efficiency. As we academics know all too well, for instance, just one malicious colleague can seriously diminish the efficiency of a department, in many respects. Meetings may take longer. Decisions may be the result of awkward compromises. Some junior department members might even be unfairly denied tenure, and drop out of academia despite having much to contribute.

Several other legitimate criticisms of Kuhn's image of science are possible—see, for instance, Mizrahi's (2015) attack on the semantic incommensurability thesis—but those I offer are sufficient to suggest what a suitable replacement might look like. In the penultimate section, which follows, I shall say something about this.

3. BEYOND KUHN'S IMAGE: RATIONAL PIECEMEAL CHANGE

In Rowbottom (2011a & 2013), I argue that science may be better served by normal science and extraordinary science—or something akin to each—coexisting to some extent. More specifically, I argue that science may benefit from having puzzle-solving functions, critical functions, and imaginative functions being performed simultaneously by its practitioners. But this doesn't mean that each and every scientist should perform all such functions.

One of the merits of the picture is that it recognizes that there are "horses for courses": it embraces the fact that some scientists are better suited to, and hence deployed in, performing some functions rather than others. For instance, a highly critical and imaginative scientist may be poor at puzzle solving: he may be too flighty, too easily bored by such activities. Nevertheless, he may be excellent at identifying previously unknown anomalies, and generating promising new theories (that others might want to investigate at some point). His presence might therefore result in the identification of anomalies more swiftly than would happen in normal science.[12] (This scientist may be well aware of what puzzle solving has been, and is being, done. This only requires keeping up with the literature in the field.)

Similarly, a highly dogmatic scientist, intent on defending her pet theories and approaches come what may, might be excellent at pushing those resources to the maximum. She might continue to try to save her favored theories when less stubborn scientists would have given up—and even when the field, as a whole, considers them falsified. And it's possible that such activity might prove beneficial. Even if there's only a small probability that it will prove successful, it's allowable, from the perspective of the group as a whole, in so far as this scientist might be useable in no other, or at least no better, way. Thus, the group may exploit her, even if her dogmatism is irrational, provided that her outputs are evaluated and employed appropriately.

So how should the balance between the different functions—and subfunctions, such as articulation and evaluation, which I will not enumerate here—be struck? This is context dependent. It depends on the scientists available as well as the state of science at the time. With respect to the latter, for instance, the amount of effort devoted to imagination may usefully rise in response to a situation where existing theories don't appear to be sufficient to solve several puzzles or problems (and usefully lower when there are few apparently insoluble puzzles). But there needn't be a crisis, or extraordinary science involving all (or almost all) scientists, or anything like that. In short, science may be highly adaptive. This is what I mean by *methodological contextualism*, which I commend to you as a more attractive (social epistemological) theory of change than Kuhn's comparatively rigid, cyclical, alternative.

I hope this makes the basic view clear. In the remainder of this section, I will build on this previous work by considering how *piecemeal change in science might rationally occur* on such a picture. I shall argue that *partial DMs and groups thereof may be used as (temporary) fixed points to enable exploration*.

There are three things I should say to prepare the ground. First, I will make free use of many of the concepts used by Kuhn and discuss many of the aspects of science that he perceptively identified. Especially noteworthy are his aforementioned ideas that dogmatism can prove useful and that exemplars are of considerable epistemic and pragmatic significance.

Second, my account is inspired by the reticulated view of change presented by Laudan (1984). The basic idea behind Laudan's view is simple: scientists may reasonably hold some parts of their framework fixed, while considering how the others might beneficially change. With this, I wholeheartedly agree. However, Laudan's view was mainly concerned with the triad of theories, methods, and aims; in essence, he argued that scientists might hold two members of the triad fixed, while exploring the wisdom of altering the other item. My account differs from Laudan's primarily, as you will shortly see, in so far as it is: more fine-grained—it involves more than theories, methods, and aims; and less restrictive—it allows rational piecemeal change to involve altering several items simultaneously.[13]

Third, I should draw brief attention to the notion of "incommensurability," which Kuhn uses to refer to the "relation of methodological, observational and conceptual disparity between paradigms [DMs]" (Sankey 1993, 759) in his early writing. On the view I'll present, such disparity may be limited such that it is never so severe as Kuhn suggests it is. More specifically, I will show how it's possible for it to be incorrect that "the transition between competing paradigms [DMs] cannot be made a step at a time" (1996, 150), and that "transfer of allegiance from paradigm to paradigm [DM to DM] is a conversion experience" (1996, 151). I take the falsity of these claims to be a *consequence* of the way that piecemeal change can occur.

This brings me to my alternative account of scientific change. The idea I want to promote is that a given group of scientists may legitimately work with a *partial* DM, or a set of *partially overlapping partial* DMs. A partial DM is just like a DM, but with some parts stripped away or diminished. So it is partial precisely in so far as it doesn't involve all the elements that Kuhn takes DMs to involve, or involves some such elements in a lesser capacity.[14] For instance, a partial DM might involve no shared theory, despite having many well-defined shared values, shared methods, and shared metaphysical commitments.[15] The absence of any shared theory doesn't entail that there aren't active theories. On the contrary, there may be several active theories. There could be considerable difference of opinion, in the community, about which theories are superior (given the values and evidence).

Partial DMs may also involve a relative paucity of shared content—be diminished by comparison with a DM, that's to say—in a particular respect. A partial DM of this kind, for instance, might involve very few shared exemplars. Many different shared exemplars might nevertheless exist in different subcommunities: there might just be widespread differences of opinion about which approaches are best, perhaps because the shared values don't suggest a unique supremacy ordering.

One needs to think about how the various components of DMs relate in order to grasp what's possible when it comes to partial DMs. That's because these components are interconnected in interesting ways: for instance, some methods and exemplars are theory-specific. But not all need be. This is clear in the case of methods, which may be experimental: for example, there might be established ways to categorize the brightness of celestial bodies without there being any shared theories about what the bodies are, how they move, and so forth. With exemplars, matters are less clear; but note that one can have the same kind of puzzle-solving processes—at *some* level of abstraction, at least—when employing different theories or symbolic generalizations. Consider two different theories about the forces exerted by a spring relative to its displacement. One is $F=-kx$, where k is a constant that may differ in value for different springs. (This is our very own Hooke's law.) The other is superficially the same, $F=-kx$, but k is a universal constant (i.e., the same for all springs). It is clear that many of puzzles about the forces (putatively) exerted by springs are going to be solved in a similar way in both cases. But the theories *aren't* the same. Something similar is true, moreover, if we allow relatively minor variation in symbolic generalizations. Consider, for instance, $F=-kx+1$.[16]

This brings me to *partial overlap* between partial DMs, which should now be easy to comprehend. Partial overlap may exist in so far as some elements may be shared. The relative extent of partial overlap is also easy to determine in some cases. For example, consider three partial DMs: A, B, and C. All components of B and all components of C are also components of A. Moreover, all components of C are also components of B. However, A has more components than B, and B has more components than C. Thus, B overlaps with A more than C overlaps with A.[17]

Consider now how partial DMs may be employed on a regular, even ongoing, basis in science. First, scientists may use them in order to explore a broader range of approaches simultaneously than would be allowed by a DM. So rather than waiting for crisis, the variety of partial DMs employed might be adjusted, on an ongoing basis, in line with the extent to which anomalies prove persistent. Moreover, much more significantly, *using multiple partial DMs may be useful for exploring possibilities that Kuhn's recommended approach doesn't allow*, and which are especially interesting from an antirealist perspective on science. For example, it may be the case that no available (or even conceivable) theory can solve all the available puzzles (in a domain), although using two or more different and conflicting available theories, in different contexts, suffices to solve the puzzles. The idea here, in short, is that scientists might allow inconsistency at the global level, and even welcome it as a better alternative than a consistent system with less puzzle-solving power.